MW00770086

WILD FOOD PLANTS

OF AUSTRALIA

TIM LOW

INCH
0
1
2
3
4
5
130
120
110
100
90
80
70
60
50
40
30
20
10 mm
0

RMM

WILD FOOD PLANTS

OF AUSTRALIA

TIM LOW

Angus&Robertson
An imprint of HarperCollins*Publishers*

To Annette

Angus&Robertson
An imprint of HarperCollins*Publishers,* Australia

First published in Australia by
Angus & Robertson Publishers in 1988

This revised edition published in 1991
by HarperCollins*Publishers* Pty Limited
ABN 36 009 913 517
A member of the HarperCollins*Publishers* (Australia) Pty Limited Group
www.harpercollins.com.au

HarperCollins*Publishers*
25 Ryde Road, Pymble, Sydney, NSW 2073, Australia
31 View Road, Glenfield, Auckland 10, New Zealand
77–85 Fulham Palace Road, London W6 8JB, United Kingdom
2 Bloor Street East, 20th floor, Toronto, Ontario M4W 1A8, Canada
10 East 53rd Street, New York, NY 10022, USA

National Library of Australia Cataloguing-in-Publication data:

Low, Tim, 1956–
Wild food plants of Australia.
[Rev. ed.].
Includes index.
ISBN 0 207 16930 6.
1. Wild plants, Edible—Australia—Identification.
2. Cookery (Wild foods). [3]. Aborigines, Australian Food.
I. Title. (Series: Australian nature fieldguide).
581.6320994

Typeset in Australia by Midland Typesetters, Maryborough, Victoria
Printed in China by Everbest Printing

18 17 16 15 10 11 12 13

Contents

PREFACE

This book is a completely revised and expanded edition of my earlier field guide, *Wild Food Plants of Australia*, published in 1988 and reprinted in 1989. Many more outback and tropical plants have been included, as well as a chapter each on mushrooms and introduced food plants.

It is impossible to produce a complete guide to wild food plants in Australia as estimates of the number of edible species run to several thousand. No guide could hope to describe and illustrate all of these. Instead, I have placed most emphasis on the foods found in the most populated area—south-eastern Australia, between Adelaide and Brisbane—and readers will find the guide to be fairly comprehensive for this region. Desert plants are also well covered, apart from the grass seeds and smaller fruits and seeds. The more common, well-known tropical plant foods are included, but not the hundreds of obscure rainforest and monsoon forest fruits that occur in eastern and northern Australia. Western Australian plants are not well covered. Plants that are edible raw or after simple preparation are given priority over those, like the grains and rainforest seeds, that require complex processing. Many species excluded from this guide for want of space are illustrated in my book *Bush Tucker*, published by Angus & Robertson in 1989.

Some wild food plants are dangerous, and the author and the publisher accept no responsibility for any mishaps arising from the tasting of plants mentioned herein.

I thank the staff of the state herbaria of Queensland, New South Wales, Victoria, South Australia and the Northern Territory for their assistance with identifications and distributions. Peter Latz of the Northern Territory Conservation Commission freely shared his knowledge of desert plants, and I have drawn heavily from his thesis. Trevor Hawkeswood supplied many of the photos, and Wendy Low shared her flat. Annette Read assisted in many ways. Finally, I thank the many Aboriginal people in Queensland and the Northern Territory who showed me plants.

TIM LOW
June 1990

1 INTRODUCTION

The Australian bush harbours a bounty of wild plant foods, ranging from tangy fruits and starchy seeds to leaves, tubers, mushrooms and seaweeds. The number of edible species runs into the thousands. Early accounts of many of these foods can be found in the journals of explorers, pioneers, colonial botanists, and early ethnographers.

Knowledge of Aboriginal diet, especially in temperate Australia, is very incomplete. The Aboriginal lifestyle was obliterated long before it could be recorded. The delicious wild parsnip, found growing near Brisbane and Sydney, was probably a major staple food, but there is not a single record of Aborigines eating it. Nor are there more than

Fruits of sweet fan flower (Scaevola calendulacea) *were no doubt eaten by Aborigines, but no record of this survives. This creeper is found on temperate beaches.*

a handful of written records of their eating rainforest fruits in New South Wales and southern Queensland, though the number of edible species runs to dozens.

It is important to realise that Aborigines did not eat everything that was edible. They had no cooking pots and could not boil foods. Most of the leaves boiled up as vegetables by convicts and settlers were not considered by Aborigines to be edible. Other foods were disregarded for less obvious reasons. Cultural preferences and lack of local knowledge probably played a part—northern Aborigines almost never ate leaves, and southern Aborigines did not eat grey mangrove starch, a staple food in the north. Beach bean was eaten by some tribes, but regarded by others as poisonous. (Other examples are cited in my *Wild Food Plants of Australia*, and in *Bushfires and Bushtucker: Aborigines and Plants in Central Australia*, a thesis by Peter Latz.)

The first known Europeans to sample wild foods in Australia were the crew of Dutch explorer Willem de Vlamingh in Western Australia in 1696. They ate raw cycad seeds and vomited so violently "there was hardly any difference between us and death". Captain Cook conquered scurvy by feeding his men numerous wild foods, including New Zealand spinach, sea celery, and magenta lilly pilly at Botany Bay, and sea purslane, taro, beach bean, cluster figs, palm hearts and Burdekin plums at Endeavour River. Of the later, overland explorers, Ludwig Leichhardt was by far the most adventurous. He gobbled many kinds of fruits, tubers and gums, also lizards, bird feet, and a sick dingo. Charles Sturt and John Stuart would probably have died of scurvy in central Australia had they not been cured by wild fruits.

The first convict settlement at Port Jackson was stalked by famine, so wild plants were used to stretch meagre rations and to treat scurvy. Desperate convicts ate seashore leaves, wild currants, figs and orchid tubers. As settlement fanned out across Australia, wild plants (and game) were relied upon to supplement dry rations. Wild fruits were baked into pies, and wild greens, especially pigweed, were boiled as vegetables. In country areas wild fruits and pigweed are still eaten today.

To the Aborigines, plant foods supplied up to 80 per cent of their diet in desert areas, and as little as 40 per cent in coastal districts, where fish, shellfish and game were abundant. The responsibility for gathering plant foods fell to the women—men snacked on fruits while hunting, but rarely brought foods back to camp. The women were consummate botanists with an intimate knowledge of the land. They could find tubers by spotting withered stems or dried stalks, and could remember the precise locations of fruiting shrubs seen months before.

Tubers were dug with long, straight, hardwood digging sticks (also used by women for fighting). Aboriginal women nowadays use straight steel crowbars—not shovels, which are considered clumsy.

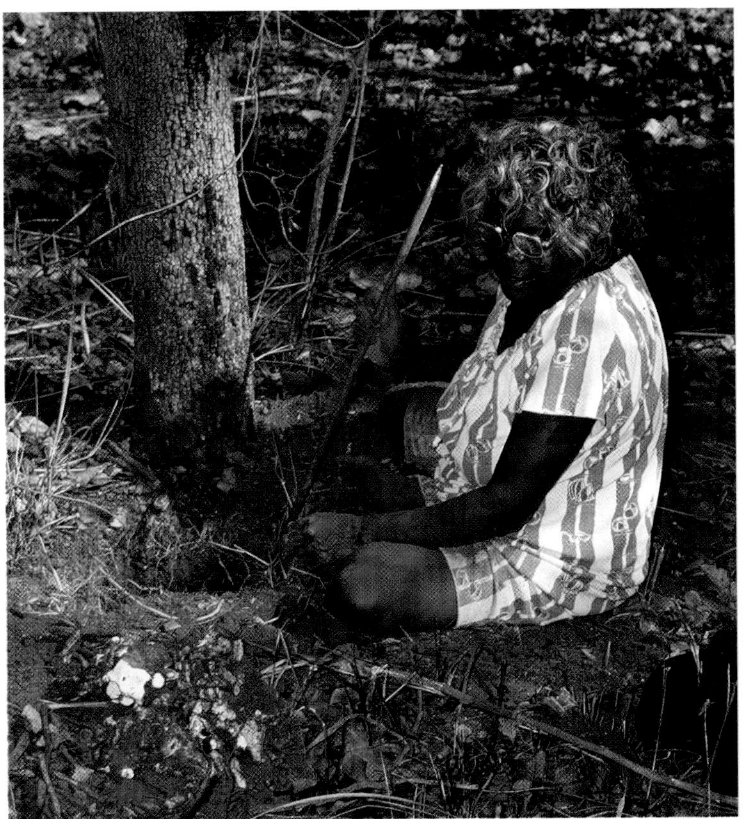

*Agnes Lippo of Belyuen near Darwin digs the tuber of a long yam (*Dioscorea*) with a steel crowbar. Pieces of the huge yam lie in the foreground and the dried stem of the yam vine can be seen twining around the tree.*

Aboriginal food foraging was so efficient that large gatherings could be supported. (The Aboriginal population of Australia has probably been underestimated, judging by the many historical references to hundreds of Aborigines camping together along rivers and coasts.) The gatherings were usually timed to the fruiting or seeding of abundant, highly calorific foods such as cycads, acacias, bunya nuts or wongi plums.

Contrary to popular belief, foods were sometimes stored—grains were hidden in caves, hollow trees, or under bark slabs. One store of 17 wooden dishes held an estimated 1000 kilograms of grain. In south-western Queensland biscuits were buried in underground caches with notched stones marking the spot. Tubers and dried fruits were also stored.

Sweet foods were very popular. In Victoria, nectars, lerps and sweet acacia gums were blended into sugary concoctions and sometimes left to ferment. In Western Australia weak alcohol was made by fermenting nectars. Alcohol was also made from pandanus fruit in the Northern Territory, and a potent "cider" was brewed from eucalyptus sap in Tasmania. Most tribes had no knowledge of alcohol, but outback tribes chewed pituri (*Duboisia hopwoodii*) and native tobaccos (*Nicotiana*) as potent drugs.

Plant foods were usually roasted in hot sand or ashes, baked in stone or underground ovens, or eaten raw. In the deserts, peppercresses were steamed in pits, and on parts of Cape York Peninsula yams were boiled in baler shells. Cycads and other poisonous seeds and tubers were grated and soaked, as well as cooked, to remove toxins. In the Northern Territory, aromatic leaves were often strewn over baking meats as flavouring herbs. Around Sydney, ants were crushed with fern starch to add flavour—the formic acid in the ants tastes like lemon juice.

In parts of Australia, Aborigines practised techniques suggestive of agriculture: in northern Australia, yam tops were often replanted after the harvest; in north Queensland yams were taken to offshore islands and planted as future food; and in central Australia dams were built and streams diverted to trigger the growth of pigweed and wild grains.

But although Aborigines practised some agricultural techniques, they were certainly not farmers. Anthropologists have often wondered why farming, practised widely in New Guinea, did not spread through the Torres Strait Islands to Queensland. Yams and sugar cane were grown on Prince of Wales Island, an island lying within sight of Australia that was visited regularly by friendly Cape York Aborigines. On more northerly islands sweet potatoes, taros, bananas, coconuts, mangoes, yams and sugar cane were grown. Some anthropologists have tried to argue that Cape York was too dry for agriculture, or that farming fades into nomadism across the islands of the Strait.

The truth is more interesting. Anthropologist David Harris found that the smaller, resource-poor islands of the Strait were dependent upon farming, while the larger islands, adjacent to rich reefs, grew food only to supplement wild supplies. On Prince of Wales Island it appears that yams were grown only in the years when wild tubers were scarce. The local islanders knew about farming and occasionally practised it, but preferred to hunt and gather.

The evidence from Torres Strait begs the question of why Aborigines did not adopt agriculture. Why should they? The farming life can be one of dull routine, a monotonous grind of back-breaking labour as new fields are cleared, weeds pulled and earth upturned. The farmer's diet is usually less varied, it is not always reliable, and the risk of infectious disease is higher. The Arnhem Land Anbarra can gather enough shellfish in two hours to sustain a person for a day—Murnong was said to be so abundant that an hour's harvest fed a family for a day.

The great gains of civilisation—the development of the arts and sciences, the diversity of lifestyles—depend among other things on an established class structure and a large surplus of productivity. These are not the realities of the simple farmer, and it is not surprising that throughout the world many cultures spurned agriculture. The Aborigines, like the Indians of the American Prairie, valued a free and easy life, and lived as contented nomads within a secure belief system.

The more sympathetic early observers knew that the Aborigines were better off without agriculture. Explorer Major Mitchell wrote in 1848:

> Such health and exemption from disease; such intensity of existence, in short, must be far beyond the enjoyments of civilised men, with all that art can do for them; and the proof of this is to be found in the failure of all attempts to persuade these free denizens of uncivilised earth to forsake it for tilled soil.

Tom Petrie, who travelled widely with the "blacks" as a boy, so learning their languages and ways, recalled in 1904:

> To them it was a real pleasure getting their food; they were so light hearted and gay, nothing troubled them; they had no bills to meet or wages to pay. And there were no missionaries in those days to make them think how bad they were.

2
Using This Book

The plant portraits of Chapter 6 are the main focus of this book. Plants have been divided by habitat into Seashore, Freshwater, Rainforest, Open Forest and Arid Zone plants. This arrangement is not as useful as it might seem because many plants range across more than one habitat. A surprising number of seashore plants can also be found growing hundreds of kilometres inland on saline soils. Currant bush (*Carissa ovata*), for example, grows on rainforest edges, in arid woodlands, and behind mangrove swamps. Placement of such plants has sometimes been arbitrary, but is mainly based upon the habitat where they are most likely to be encountered.

Plants or groups of plants with very wide habitat tolerances, like mistletoes, orchids and banksias, are placed in the Open Forest section—a large and miscellaneous section which should always be perused when identifying plants.

The habitats and their characteristics are described in the next chapter.

Names

Where scientists disagree about the correct scientific name for a plant, the less popular alternative is listed under "Other names" without brackets. Old invalid scientific names, if still being used, are listed in square brackets. Many scientific names have been changed in recent years and the reader should expect discrepancies in older books or on nursery labels (for example, *Eugenia coolminiana* has become *Syzygium oleosum*).

There were hundreds of languages in Aboriginal Australia and it has not been possible to list Aboriginal plant names, except for those incorporated into English (for example, murnong, midyim, muntari, geebung). Another exception has been allowed for desert foods, where the main Pitjantjatjara (abbreviated as P) and Arrante (abbreviated as A) names are given. These are included because of the increasing interest shown in wild foods in central Australia. Arrante is the main language spoken at Alice Springs and Pitjantjatjara at Uluru (Ayers Rock).

The early pioneers often named their wild fruits after vaguely similar cultivated foods, as in "native cherry" or "wild orange". In most cases the native species are unrelated to their namesakes. The exceptions are the native figs, raspberries, elderberries, bananas, passionfruits and melon, which are true figs, raspberries, elderberries and so forth. Also, the native grapes, tomatoes, limes and cashew are members of the same plant families as their namesakes.

DISTRIBUTION

Plant distributions are shown on maps. Where related species are described together (for example, apple-berries, pigfaces) the maps show the combined range of all the edible species. The distributions of rainforest and coastal plants should be nearly accurate, but the ranges of many desert plants are poorly known. The big gaps in distribution shown on some maps (for example, chocolate lilies, black plum) reflect Australia's increasing aridity, and the restriction of species to wet pockets separated by dry stretches.

IDENTIFICATION

Most plants should be recognisable from the illustrations alone, bearing in mind differences in flower and fruit colours. The length and intricacy of the description should be taken as a guide to the likelihood of misidentification. Descriptions of fruit-bearing plants may be inadequate to identify plants not in fruit. Fruit colours and measurements assume the fruit is ripe—unripe fruits are usually hard, green and very distasteful. Leaf measurements do not include the stalk.

This guide is intended for the non-botanist and botanical jargon has been avoided. The words "flower", "petal", "fruit", "seed", "leaf" and "tuber" are given their colloquial meanings.

The following semi-technical terms are used in the text.

Fruit Features

Aromatic: having a strong, sometimes resinous flavour that would be fragrant in smaller doses.
Astringent: producing a drying, puckering sensation in the mouth when the fruit or leaf is chewed. The tongue and teeth feel "fuzzy". Banana skins and unripe persimmons are astringent.

Bitter: the taste of lemon seeds, very old lettuce leaves or bitter beer.
Sour: the acid taste of lemon juice or vinegar.

Leaf Features

Opposite leaves: the leaves are produced along the stems in opposing pairs.
Alternate leaves: the leaves are produced singly, alternating along the stems.
Pinnate leaves: each leaf is divided into a series of smaller segments, like a fern frond, palm, or rose leaf.
Leaflet: the individual leaf segment of a pinnate leaf.

As points of reference, eucalypt leaves can be considered *tough* and *leathery*, lettuce leaves are *thin*, and tomato leaves are *soft* and *furry* (or *hairy*).

Uses

Not all of Australia's edible plants were eaten by the Aborigines. Seablite and coastal saltbush, for example, though widely used by colonists, were apparently disregarded by Aborigines. If a plant is a recorded Aboriginal food this is nearly always listed in the text.

3

THE HABITATS

The plants in this book have been ordered by habitat. What follows is a brief description of each of these.

SEASHORES

The seashore environment includes coastal dunes, rocky headlands, saltmarshes and mangrove swamps. Seashores were very rich environments for Aborigines, yielding plenty of plant and animal foods, including shellfish, crabs, mangrove worms, fish, turtles, seals and dugongs.

The seashore environment is particularly rich in edible fruits and leaves. Tubers are scarce or absent on temperate coasts, but are common behind beaches in the tropics, where long yams, Polynesian arrowroot, convolvuluses and tar vines can be found. Seeds were also important in the tropics, especially the pandanus, mangrove pods, sea almond, and introduced coconut.

Sea celery (Apium prostratum) *is a common herb of southern Australian beaches.*

When identifying southern seashore fruits, refer also to the Open Forest section (for devil's twines, flax lilies, heaths, banksias, and native cherries), and the Arid Zone section (for nitre bush, ruby saltbush and quandong). In the tropics refer to the Open Forest zone for sandpaper figs, white berry bush, lady apple and broad-leaved native cherry. In eastern Australia rainforests often grew by the sea, and in the surviving patches typical rainforest fruits such as lilly pillies and figs can be found. Of the freshwater plants the sea club-rush, water ribbons, bungwall and native lasiandra often grow close to the shore.

Freshwater

Freshwater plants include those living in or at the edge of rivers, streams, lakes, swamps, claypans, and seasonally inundated hollows. Many freshwater plants are subject to extreme changes in water levels, and they survive this instability by producing starch- and water-filled tubers to tide them through dry spells. The freshwater environment is exceptionally rich in tuberous plants, and provided a disproportionate share of Aboriginal diet.

In the outback, wild foods often grow prolifically on river banks and claypans after flooding. Nardoo, pigweed, native millet, nalgoo and ulcardo melons often sprout in lush fields. In the drier regions fruiting shrubs and trees such as boobialla, sandalwood and wild orange are restricted to the banks of watercourses.

Many rainforest trees such as Moreton Bay chestnuts and sandpaper figs are largely restricted to river banks. In north-western Australia scrubs rich in yams and fruits are found along sheltered creeks.

Rainforest

In rainforests trees are closely spaced, the forest floor is shaded, vines and epiphytes are common, and grasses and herbs are rare.

Botanists recognise many kinds of rainforests. Best known are the tropical and subtropical rainforests of coastal eastern Australia, found south to about Nowra. The very diverse flora includes hundreds of different trees and vines, as well as ferns and palms. Both tropical and subtropical rainforests are very rich in bush tucker, especially fruits.

By contrast, the temperate rainforests of Tasmania, Victoria and the New South Wales tablelands contain little to eat apart from tree ferns and a few fruits. These mossy forests can be distinguished from subtropical rainforests by their low diversity, and lack of palms, strangler figs, large vines and buttressed roots.

The so-called "dry rainforests", including dry scrubs and vine scrubs, grow where the rainfall is lower, especially west of the Dividing Range.

In these rainforests, the canopy is lower (especially in the scrubs), mosses and ferns are rare, and prickly shrubs are common. Dry rainforests are rich in edible fruits, including many obscure species not included in this book.

In far northern Australia, on the better soils or along creeks, grow patches of monsoon rainforest. These rainforests have a low canopy and lack ferns or mosses. Many of the trees are deciduous, losing their leaves during the harsh winter-spring dry season. Monsoon rainforests are very rich in edible fruits and tubers. The Northern Territory monsoon forests, for example, contain 40 species of edible fruit. The tuber bearers include the long yam, round yam and Polynesian arrowroot.

Rainforest foods are difficult to present in field guide format because there are too many fruits—several hundred species in all. I have concentrated on the subtropical species of New South Wales and southern Queensland. Only the more common and spectacular tropical fruits have been included, whilst the fruits of both dry and monsoon rainforests have been largely ignored. John Brock's *Top End Native Plants* lists and illustrates the monsoon species, and is strongly recommended reading.

Of the excluded fruits, the native bananas and native passionfruits deserve mention for their affinity to known crops. The native banana (*Musa acuminata*), found in north Queensland (and common at Cape Tribulation), resembles the cultivated banana but has fruits filled with black seeds; the native passionfruits (*Passiflora*), found in eastern Australia, have three-lobed leaves and soft-skinned egg-shaped fruits.

OPEN FORESTS

Australia's open forests are the eucalypt forests of coastal and near-coastal regions. In temperate regions this typically Australian bush consists of a loose canopy of eucalypts and related trees, above an understorey of shrubs (mainly wattles and native peas) and a ground cover of grasses and herbs. In northern Australia the eucalypts share the canopy with many fruit-bearing trees.

Temperate open forests are often lacking in wild foods. Apart from geebungs, native cherries, acacias, cycads and occasional heaths, the trees and shrubs produce little that is edible. Where cattle and sheep have not grazed, the herb layer may yield tuberous lilies, orchids and murnong, as well as bracken and cranesbills.

In the tropics, there are many large-leaved trees producing edible fruits, notably the cocky apple, lady apple, green plum, and billygoat plum. Yams, Polynesian arrowroot and native grapes produce edible tubers.

HEATHS

Heaths are communities of hard-leaved shrubs and small trees found on infertile sandy soils in eastern and south-western Australia. The same name is given to shrubs in the family Epacridaceae, which are common in these habitats. Heathland plants are included in this book under the Open Forest section, as many heathland plants grow in both habitats.

Heaths are very rich in wild foods, especially small fruits, though most are snack foods rather than staples. Typical heath fruits include the native cherries, heaths, devil's twines, geebungs and currant bushes. Banksias, grasstrees, ground orchids, lilies and the native parsnip are other heath plants of culinary significance.

Alpine heaths are a feature of high altitudes in southern Australia. The wild foods include ground orchids (especially *Prasophyllum alpinum, P. suttonii*), and many small red fruits (see page 18).

ARID ZONE

Australia's outback is an enormous, diverse region embracing many different habitats, each with its own community of plants. The edible species tend to be widespread and found in many situations. Apart from the plants listed under Arid Zone, refer to the Seashore section for New Zealand spinach, pigface and muntari, to the Freshwater section for nardoo, and to the Open Forest section for wood sorrel, sowthistle, lilies, mistletoes, lerp and acacia gum, among others.

Desert Aborigines ate many more seeds than coastal tribes, especially grass, acacia and pigweed seeds, and tiny spores of nardoo. Leaves were also more widely eaten. Fruits were very important, and quandong and wild tomatoes were sometimes dried for future use. Only a few kinds of tubers occur in the outback, and they are completely absent from some deserts. Desert Aborigines ate more plant foods than coastal groups.

Photos taken in the 1950s and 1960s of the last traditional Aborigines living in the deserts convey the idea that outback life was exceptionally harsh. This was not always so. Tribes lucky enough to control the rivers thrived on a rich diet of fish, water birds and seeds.

4
The Plants as Foods

Most plants are toxic and indigestible to humans. The exceptions, the food plants, fall into one of three categories: they may be plant parts that are designed to be eaten (nectars and fruits); or parts that are unprotected by fibres or toxins (most tubers and small seeds); or they may have chemical defences that can be removed by leaching or cooking (many large tubers and seeds).

To understand why some plants are edible and others are poisonous it is necessary to look at the ecology and lifestyle of plants. This chapter briefly examines some of the principles involved.

Flowers

Flowers are the sex organs of plants and their role is to exchange pollen. Many flowers, like those of grasses, plantains and pine trees, are pollinated by wind. Small and dull looking, they produce no nectar, and have no food value.

Animal-pollinated flowers are larger, more colourful, and usually contain stores of sweet nectar. The largest flowers are those pollinated by birds and mammals such as honeyeaters, lorikeets, fruit bats, and small possums. Flowers dependent on large pollinators need to provide plenty of nectar, and this makes them a good source of food for humans too. Australia is very rich in vertebrate-pollinated flowers. Examples include the banksias, many grevilleas and tea trees (*Melaleuca*), some eucalypts and hakeas (*Hakea*), the waratah (*Telopea speciosissima*), the honey lambertia (*Lambertia formosa*), certain fuchsia bushes (*Eremophila*), grasstrees (*Xanthorrhoea*), and wild bauhinias (*Lysiphyllum gilvum, L. carronii*). All but the last four of these are members of Australia's two most characteristic plant families, the Proteaceae and Myrtaceae.

Apart from honey, honeypot ants and perhaps manna, nectars were the sweetest foods available to Aborigines and were highly prized. Drinks made by soaking banksia, grasstree and bauhinia blossoms in water were sometimes left to ferment into a weak mead.

Aborigines loved to suck nectar and to soak blossoms in water to make drinks, but they rarely ate whole flowers. The few exceptions include doubah, purple star, mat-rushes, honey pots (*Acrotriche serrulata*), native rosella buds, and kapok bush (*Cochlospermum fraseri*) petals. Petals are mere decorations to attract pollinators, and one would not expect them to be tasty or nutritious. It is not surprising that Aborigines rarely used them.

Nectars and petals are usually safe to sample, but poisonous species exist. In south-western Australia the poison bushes (*Oxylobium*) have very toxic pea flowers, and in eastern Australia the crinkle bush (*Lomatia silaifolia*) is thought to have poisonous nectar. The garden oleander also has very toxic nectar.

Seeds

The germination of a seed in the soil is a critical time in the life of a plant and most seeds come equipped with a store of food in the form of starches, proteins and fats. These nutrients make them very valuable as human foods, and indeed, seeds are the most concentrated and energy rich of all plant foods. They form the staple diet of most cultures.

Of the seeds eaten by coastal Aborigines, many were large and poisonous. Very big seeds are found on rainforest trees, cycads, and plants of tropical seashores. These big seeds can be a liability to a

Bunya nuts are very nourishing. They can easily be gathered from trees planted in parks.

plant, for they are not easily dispersed, and rats are strongly attracted to them. But their large food reserves are helpful to seedlings germinating in difficult environments such as gloomy rainforests or salty mudflats.

Along seashores the large seeds include the coconut (Australia's biggest seed), matchbox bean (*Entada phaseoloides*), beach bean and certain mangroves, all of which are dispersed by water.

In the rainforests there is the bunya nut, macadamia, and monkey nut, among others. Macadamias and candlenuts are rich in oil, and have very hard shells to deter rats from stealing this precious food.

Many plants with large starchy seeds, particularly the cycads, Moreton Bay chestnut, matchbox bean, and several Queensland rainforest nuts, deter rats by producing seeds laced with poisons. Aborigines were very adept at grating or pounding, soaking, and baking poisonous seeds to remove the threat. The starch so produced served as a staple food.

Small seeds are usually poison-free, but only a few kinds were eaten by coastal Aborigines, notably a few wattles, native beans, and native flax (*Linum marginale*).

In the outback, Aborigines ate many kinds of seeds, most of them very small. The largest were quandong kernels, followed by kurrajong seeds, acacias, grass seeds, pigweed and nardoo.

Grass seeds were very important outback staples, especially woollybutt and native millet. It is noteworthy that apart from wild rice (*Oryza meridionalis*), eaten in the Gulf of Carpentaria, grass seeds were not eaten by coastal Aborigines, even though their use in western New South Wales dates back 15,000 years. It has been suggested that outback Aborigines only turned to grains after diprotodons and other large game became extinct.

The gathering of wild grain was very organised. The explorer A. C. Gregory described a harvest on Cooper's Creek:

> Fields of 1,000 acres are there met with growing this cereal. The natives cut it down by means of stone knives, cutting down the stalk halfway, beat out the seeds, leaving the straw which is often met with in large heaps; they winnow by tossing seed and husk into the air, the wind carrying away the husks. The grinding into meal is done by means of two stones, a large irregular slab and a small cannon-ball-like one.

Australia has more than 850 different native beans and peas, but nearly all are poisonous, and only a few were harvested by Aborigines. These were mainly tropical species with very big seeds which were worth the effort of leaching and special preparation to remove the poisons. Beans and peas are legumes, and as such use the bacteria in their roots to draw nitrogen from the air. Nitrogen is needed to produce protein and to manufacture toxins. Australia's soils are notoriously nitrogen-deficient, and most Australian plants cannot afford to divert nitrogen from growth into toxin production. Because legumes have free access to nitrogen, they face no such problem, and for this reason their seeds are invariably well protected with poisons.

FRUITS

Fruits, like flowers, can be very alluring—they beckon with dazzling colours and shiny shapes. They look enticing because, like the nectar in flowers, they are designed to be eaten. When an animal swallows a fruit the seeds usually pass through unharmed, to be voided far away, and in this way fruit-bearing plants disperse their seeds.

Some animals are more effective at dispersing seeds than others. Birds are excellent dispersal agents because they usually swallow fruits whole, and may fly many kilometres before voiding the seeds. Mammals on the other hand, are usually poor dispersal agents because they use teeth and claws to chew and destroy seeds or discard them on the spot. Fruit bats, or flying foxes, are the exceptions among Australian mammals in being very good at dispersing seeds.

Over millions of years, close bonds have evolved between fruiting

Birds disperse the sweet fruits of lollyberry (Salacia chinensis), *a twiner of tropical beaches.*

plants and the animals that best disperse their seeds. Most Australian fruits have developed characteristics to make them attractive either to birds or fruit bats. The bewildering kaleidoscope of fruit colours, shapes and smells demonstrates this.

Birds see colours, but (with few exceptions) cannot smell. Fruit bats are colour blind, but have a keen sense of smell. Bird fruits tend to be small, brightly coloured but not odorous, whilst bat fruits are usually sombre, smelly and large.

Fruit bats are largely restricted to northern and eastern Australia, so bat fruits grow only in these regions. The most southerly bat fruits are certain figs and the black plum, which extend to southern New South Wales.

Bat fruits are often large, with soft or stringy pulp, a heavy musky odour, and a tropical fruit flavour. They are nearly all edible to people. Most cultivated tropical fruits and some stone fruits were developed from bat fruits—notably the mango, pawpaw, banana, guava, custard apple, peach, apricot, and date. Many native bat fruits, like the billygoat plum, great morinda and Leichhardt tree fruit, are high in vitamin C.

Bats hang downwards when they feed, and many bat fruits hang with them, including the pandanus, Davidson's plum, native bananas, and cultivated mango and pawpaw. Some bat fruits—the larger figs and lilly pillies—sprout directly from the trunk, beyond reach of fruit pigeons, which cannot hover or perch there.

Bats are clumsy fliers, so the trees they feed in often have an open canopy for easy access, as in the pandanus, figs, cultivated mango, pawpaw and banana. Some bat fruit trees, like the Leichhardt tree and sea almond, have evolved a "pagoda" structure of horizontal tiered branches with open spaces between. Pagoda trees are widely planted in Asian cities for their pleasing shape—a gesture much appreciated by the bats.

Bats quarrel when they eat, squawking and feeding like seagulls in the night. To escape the conflict each bat may fly off with a fruit in its mouth to feed elsewhere in peace. Seeds are in this way dispersed to nearby feeding trees, or are dropped accidentally en route. (In Queensland, bats often drop mangoes onto tin roofs, with startling effect.) Bats do not swallow large seeds.

Fruit bats do, however, feed on many kinds of fruits (such as small lilly pillies) to which they are not obviously adapted, and there are some fruits, notably the smaller figs, which seem to be both bat and bird adapted.

Bird fruits vary enormously, as there are so many different fruit-eating birds, ranging from tiny silver-eyes and honeyeaters, through pigeons, bowerbirds and orioles, to scrub turkeys, bustards, the emu and cassowary. Most Australian fruits are bird fruits.

The native cherries (*Exocarpos*), elderberries and mistletoes are typical of fruits eaten by small birds. They are bite-size for a honeyeater or mistletoe bird, and are carried on slender twigs beyond reach of larger animals. Nearly all fruits produced on shrubs, creepers and herbs, or found growing in central or southern Australia, are bird fruits.

In the outback there are several fruits obviously designed for emus, notably the quandong, emu apple, wild tomatoes and nitre bush. Nitre fruits rarely germinate unless they have passed through an emu's gut. The bird's stomach strips a layer off the seed that inhibits germination.

Similarly, the north Queensland rainforests hold many large and colourful fruits designed for dispersal by cassowaries. The red bopple nut of northern New South Wales appears to be a cassowary fruit, although cassowaries have been extinct in this region for hundreds or thousands of years (which may explain why this tree is rare). The rainforests also contain many pigeon fruits such as the blue quandong and crab apple.

In the Australian Alps nearly all the fruits are small, red, bird fruits. Unrelated plants like the alpine heaths, alpine raspberry, perching lily (*Astelia alpina*) and mountain plum (*Podocarpus lawrencei*) have evolved remarkably similar shiny red fruits, presumably to make it easier for the few fruit-eating birds to find them.

Bird fruits are sometimes poisonous to mammals including ourselves. The toxins are probably to deter rats, which destroy enormous numbers of forest seeds. In a survival situation it is helpful to distinguish bird and bat fruits; the latter are much less likely to be poisonous.

Apart from bird and bat fruits, there are a number of fruits which appear to be designed for dispersal by other mammals, or perhaps by reptiles. The bolwarra, cranberry heath, and most geebungs have fruits that are green and inconspicuous to birds. Geebungs and bolwarra fruits fall to the ground when ripe, implying that their dispersal agent is a ground-dwelling animal. Geebungs have very hard stones and are known to be dispersed by kangaroos, although birds also eat them. The very large, strong-smelling fruits of the inland capers may be targeted at diprotodons and other extinct giant marsupials. Many heaths have tiny green fruits and their dispersal agents remain a mystery.

Leaves

Plants draw their energy directly from sunshine, which they trap in solar discs called leaves. The process uses the green pigment chlorophyll and the gas carbon dioxide.

To draw in carbon dioxide, leaves need to be thin and porous; to trap sunshine they need to be flattened and prominently displayed. Because of these constraints leaves are more vulnerable to animal attack than underground roots, woody trunks or hard seeds. Consequently, most plants protect their leaves by impregnating them with tough fibres, and with toxic and distasteful chemicals.

Humans are generalised feeders with unspecialised teeth and we cannot tackle the tough fibrous leaves of trees, shrubs and grasses. Cooking softens plant tissues but doesn't help us digest cellulose. As unspecialised feeders we also lack immunity to the arsenal of poisons with which plants protect their leaves. Consequently, there are not many kinds of wild leaves that we can eat. Most can be categorised either as quick-growing herbs or halophytes.

Quick-growing herbs sprout after rain and channel nearly all their resources into growth and reproduction. Little energy is diverted to production of poisons and tough leaves. The aim is to grow quickly and set seed before being found by a plant-eater. Native peppercresses, yellow cress, pigweed, parakeelya, many garden weeds and most cultivated vegetables live like this.

Halophytes are plants adapted to saline soils, and are found mainly on beaches, saltmarshes, and inland saltbush plains. Though too much salt is fatal to plants, which cannot avoid absorbing whatever is present in the soil, halophytes get by either by excreting excess salt on to the leaf surfaces (as in the bladder cells of saltbushes), or by absorbing extra water to maintain a constant salt:water ratio, as succulent plants do. Salt is unpalatable to most plant eaters, and most halophytes rely on their saltiness for chemical defence. But this salt is easily removed by boiling, and most halophytes can be boiled and eaten as vegetables.

Explorers and settlers ate many kinds of halophytes, especially saltbushes (*Atriplex, Rhagodia, Enchylaena, Einadia*), samphires, and wallaby bush (*Threlkeldia diffusa*). Aborigines had no pots for boiling and made little use of such plants, apart from pigfaces, which were eaten with meats to make them salty. Their techniques of steaming leaves would not have removed much salt.

A few shrubs and trees of tropical beaches have young leaves that can be cooked and eaten. These plants are featured in Cribb and Cribb's *Plantlife of the Great Barrier Reef*.

Aborigines, especially in northern Australia, ate very few if any leaves, presumably because better foods were available.

STEMS

The leaves of palms, tree ferns and grasstrees emerge from a woody trunk, in the centre of which lies a core of starch—the undifferentiated tissues from which the frond is made. Aborigines tore the trunks open to extract this highly nutritious morsel.

The flame lily produces enormous flowering stalks which Aborigines harvested while still young. They also chewed the starch-filled stem bases of epiphytic orchids.

Most parallel-veined leaves (Monocots) produce leaves from a single point, and in many sedges and rushes the soft growing cluster of leaf bases is tender and substantial enough to be nibbled. Bulrushes, mat-rush, and grasstrees can be harvested in this way.

The stems of rainforest vines are not edible, but are a useful source of emergency water. Cut neatly into metre lengths, they can be drained of their watery sap into a bucket. Vines yield copious water because their stems are supported by adjacent trees and do not need to lay down supportive woody tissue. Most of the stem is a tube for conveying water to the leaves, and the yield is consequently very generous. Native grape vines were known to colonists as "water vines" for this reason. The smaller vines with rubbery stems are the better water sources; older lianas become woody.

Roots

Plants depend on their roots to provide water and minerals, and to anchor them in place. Many plants also use roots for storage. For plants that lack woody trunks, roots are the ideal places to store extra food

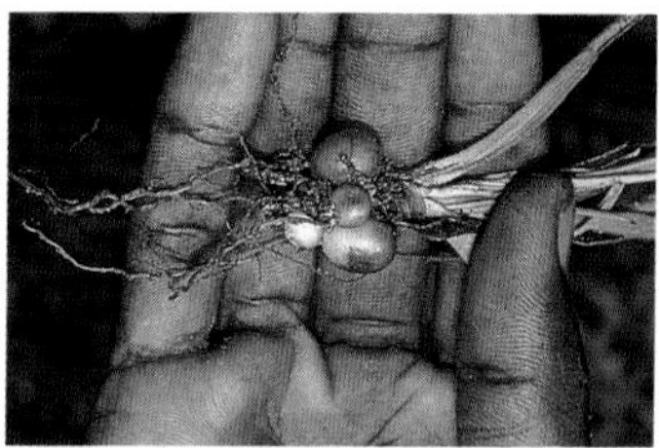

Tubers of bulbine lily (Bulbine bulbosa; *left*), *arda* (Cartonema parviflorum; *top right*) *and murnong* (*bottom right*) *are borne abundantly and could serve as important foods.*

and water, safe from animals and temperature extremes. Aborigines were adept at locating these hidden stores, and many roots served as staple foods.

Fewer than five per cent of Australian plants produce tubers. The tuberous organs range from swollen roots (root tubers) and underground stems (rhizomes or corms) to swollen overlapping leaf bases (bulbs). Tubers are common in some habitats, among certain groups of plants. In Australia the main tuber-producers are lilies, ground orchids, vines and freshwater plants.

Tuberous lilies and ground orchids are a characteristic feature of Australian woodlands. The plants spend most of the year in dormancy, as tiny underground tubers. When conditions are favourable they send up leaves, flower, set seed, then wither and revert back to tubers for the remainder of the year. In northern Australia the lilies and orchids sprout during the autumn wet season and thus avoid the hot dry spring; in southern Australia they mostly flower during spring. In the outback the few lilies flower opportunistically after rain.

The murnong or yam daisy has the same growth pattern. It sprouts leaves in autumn, begins flowering in late spring, and withers by late summer. It often grows in dense colonies alongside lilies and orchids, and Aborigines reaped a bounty of mixed tubers by turning the soil with sticks.

Freshwater habitats support many tuber-producing plants. Aboriginal staples included spike rush, bulrushes and nalgoo. The freshwater environment is very unstable—water levels rise and fall—and many plants survive the dry spells by producing tubers that lie dormant in the mud.

Many vines and creepers also produce tubers as a means of surviving harsh times. Northern Australian yams sprout foliage during the wet season and, like the lilies, die back to tubers as the hot dry winter advances. In the deserts, tuberous tar vines, bush potatoes and maloga beans sprout after good falls of rain.

Apart from the large-leaved lilies (cunjevoi, Polynesian arrowroot, etc.), relatively few of Australia's tubers are toxic. Indeed, the tubers of all starchy freshwater plants (except taro and some waterlilies), and of all 700 Australian ground orchids, appear to be edible raw.

5
The Danger of Poisoning

Most wild plants are inedible, and become poisonous if taken in large enough doses. Gum leaves, for example, are laced with irritant aromatic oils and other toxins. They do not cause poisonings because like most plants, they taste unpleasant and so are not eaten.

Poisonous plants nearly always warn of their toxicity by tasting bitter or acrid. The human taste buds are highly sophisticated and can easily detect most toxins. Victims of poisonings are usually small children whose taste buds are less discriminating.

There are, however, a few poisons that our senses of taste and smell cannot detect, and plants producing these can pose a danger. These plants are described here in detail. Poisonous mushrooms are considered in Chapter 8.

Cycads

Many explorers were poisoned by cycads (see page 139). The boiled or baked seeds taste delicious, but contain a potent toxin that must first be removed by leaching. Even the leaves of cycads are dangerous. Explorer J. M. Stuart told how one of his men in 1861 ate the lower end of a leaf and found it tasted like sugar cane: " he ate a few inches of it, and in a few hours became very sick and vomited a good deal during the evening."

Legumes

Not much is known about Australia's native beans and peas, but it seems likely that most are poisonous, and that some are deceptively tasty. Moreton Bay chestnut seeds, for example (see page 94), are very tasty when boiled, and even palatable raw, but in both conditions are poisonous. Raw beach beans and probably many of the smaller beans are not distasteful raw. The forager should be wary of sampling unknown legumes. Most species belong in the family Fabaceae and have easily recognised pea flowers. Cassias (*Cassia*) in the related Mimosaceae family are also a problem. The seeds and sweet pulp are purgative. Cassias have showy yellow flowers and bean-like pods.

TIE BUSH

Tie bush (*Wikstroemia indica*) is considered poisonous in India, and a Queensland child died, supposedly from eating the fruits. These have a bitter aftertaste but are sometimes not unpalatable. Tie bush is a small shrub of seashores, shady gullies and dry rainforests north of Wollongong. Leaves are soft and thin, 2-6 cm long, the bark tears in long strips, the small flowers are greenish-yellow with four petals, and the fruits are orange-red, 5-15 mm long, containing one seed. I have eaten one or two of the fruits without ill-effect but this is a plant that should be avoided.

Tie bush (Wikstroemia indica) *fruits turn tomato red when ripe. The leaves are poisonous to cattle.*

FINGER CHERRY

Also known as native loquat (*Rhodomyrtus macrocarpa*), this north Queensland rainforest plant is notorious for having inflicted permanent blindness. It is a shrub or small tree with blunt-tipped opposite leaves up to 25 cm long, pink or white flowers with five petals, and reddish egg-shaped fruits 2-5 cm long. The fruit has five small flaps at the end like a lilly pilly, but differs in containing a number of kidney-shaped seeds. Finger cherry is a traditional Aboriginal food, but many white people have suffered permanent blindness after eating large amounts of the fruit. It is thought that an infesting fungus causes the blindness. Foragers should avoid this plant.

LARGE LILIES

The cunjevoi, taro, black arum lily (*Typhonium brownii*), Polynesian arrowroot and related large-leaved lilies contain toxins in their roots, and often in their leaves and stems, which cause a burning pain in the mouth and throat if chewed. The onset of pain can take up to half a minute or so, which makes these plants very dangerous if large amounts are chewed and swallowed quickly. Foragers should be very wary of these plants. They all bear tubers which Aborigines ate after careful preparation.

6
THE PLANTS

SEAWEEDS

FIELD NOTES: Seaweeds come in most colours and shapes and display an enormous variety of form and expression. They flourish on cool rocky coasts, but also grow in mangrove swamps, on coral reefs, and in the open sea.

USES: Seaweeds, like insects, are a nutritious and abundant food source found worldwide, but ignored by most cultures. The Japanese, Koreans and Hawaiians are seaweed connoisseurs, using dozens of different kinds.

The only seaweed recorded in Aboriginal diet was bull kelp (*Durvillea potatorum*), eaten in Tasmania. The French explorer Labillardiere noted in 1800: "On the same fires we observed them broil that species of sea-wrack [bull kelp] . . . and when it softened to a certain point, they tore it to pieces to eat it." Another observer wrote that it was "eagerly looked for and greedily eaten, after having undergone a process of roasting and maceration in fresh water, followed by a second roasting, when, though tough . . . it is susceptible of mastication". South Australian Aborigines wrapped seaweeds around meats steamed in fires.

The carbohydrates in seaweeds are mostly indigestible, but they are rich in iron, iodine, calcium, carotene and B vitamins. Some seaweeds are culled commercially to make agar, stabilising agents, and the imitation cherries in dried fruit mixtures. Jellies can be made from red seaweeds, as A. H. S. Lucas recorded in 1936: "Ladies in Western Australia make jellies of *Euchema* and ladies in Tasmania make jellies of *Gracillaria*, jellies much appreciated at their tables." As a rule, any seaweed that does not taste bitter, spicy or irritating is safe to eat, raw or cooked.

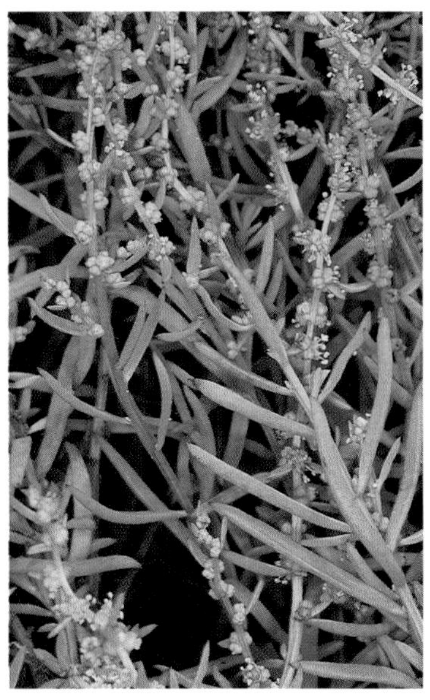

Seablite

Suaeda australis

Field Notes: The fleshy leaves are green or red, 1-4 cm long, and the tiny flowers are greenish, on stalks 15-30 cm tall. Seablite is a shrubby herb of saltflats, inland saltpans, and sandbanks behind mangroves.

Uses: Settlers used the leaves as a vegetable and pickle.

Sea Purslane

Sesuvium portulacastrum

Field Notes: Leaves are fleshy, 2-9 cm long, stems are usually reddish, and the pink starry flowers have five petals. This shrubby herb grows on beaches and mudflats.

Uses: Captain Cook ate the leaves at Endeavour River.

SAMPHIRE

Sarcocornia quinqueflora

OTHER NAMES: Chicken claws, beaded glasswort, [*Salicornia australis*]

FIELD NOTES: The succulent, jointed, leafless stems are green, reddish or purple, about 4 mm thick, growing 10–20 (rarely 50) cm tall.

Samphire forms extensive mats on tidal mudflats, where it is usually the dominant plant. One subspecies grows on rocky headlands in Tasmania and Victoria.

USES: The earliest years of convict settlement at Port Jackson were shadowed by the spectre of starvation. The soil was infertile, the rations inadequate, and scurvy ran rife. Desperate convicts and officers scoured the countryside for wild vegetables. The diary of naval officer W. Bradley records "wild spinage, samphire & parsley & small quantity of sorel and wild celery". It is no surprise that samphire appears on this list, for the Australian species looks much like its English equivalent (*S. stricta*). English samphire is today used mainly as a pickle, and Australian samphire can be used in its place, although the woody core within each stem detracts.

Samphire (or something similar) was an emergency food of explorers George Grey and Colonel Warburton in north-western Australia, and was pickled by Tasmanian colonists. *Halosarcia indica*, a tidal mudflat plant with similar jointed stems, but which grows into a sustantial shrub up to 2 m tall, was pickled by New South Wales colonists.

Sea Celery

Apium prostratum

Field Notes: This is a very variable herb with toothed or jagged pinnate leaves, tiny white flowers in clusters, and a celery smell when crushed. There are two very different forms. Plants growing on sunny coastal dunes (var. *prostratum*, illustrated page 9) are small and squat, with thick, shiny, broad leaves. Plants growing further inland near swamps or on muddy river banks (var. *filiforme*, illustrated above) have soft, thin, slender, jagged leaves and long trailing ribbed stems. Plants intermediate between these two forms are also found.

Uses: Sea celery was a significant vegetable of Australia's first explorers and colonists. It was first eaten by Captain Cook at Botany Bay, and later by Labillardiere in Tasmania in 1792. The journals of no fewer than four First Fleet Officers refer to its use as a vegetable (calling it "parsley" or "celery") during Sydney's initial starving years. In Tasmania in the 1830s botanist Daniel Bunce said it "forms an excellent ingredient in soup, and otherwise may be used as a pot herb". Around Albany in Western Australia it was even cultivated by colonists as a vegetable. Variety *filiforme* is the more palatable, tasting much like cultivated celery leaves, and was probably the main form used. Both the leaves and stems are eaten.

The related dwarf celery (*A. annum*), a tiny herb of southern beaches, is also edible, and there is a giant native celery (*A. insulare*) on Bass Strait islands and Lord Howe which could be worth investigating.

NEW ZEALAND SPINACH

Tetragonia tetragonoides

OTHER NAMES: Warrigal greens, Botany Bay spinach

FIELD NOTES: Leaves are thick and triangular, 2-12 cm long, bright green and glistening as if covered in dew or fine sugar crystals, especially on the undersides. Yellow flowers are followed by hard pods, 1-1.5 cm long, with small horns.

This is a common shrubby herb of sheltered beaches, saltmarshes, arid woodlands and plains, extending into central Australia. It has escaped from cultivation overseas to become a feral plant in Africa, Europe and the US.

USES: New Zealand spinach was one of Captain Cook's many famous discoveries. First sighted along the coasts of New Zealand, and later at Botany Bay, it was cooked and eaten by the *Endeavour* crew to allay scurvy. Joseph Banks considered it to "eat as well as spinage or very near it" and he took seeds to Kew Gardens. During the 1880s New Zealand spinach was promoted in European and American seed catalogues as a hardy, summer-growing spinach substitute. It was popular for a time, then faded into obscurity. It has recently been rediscovered by wild food promoters in Australia, and can be ordered in bush tucker restaurants as "Warrigal greens". It is the only Australian plant to be cultivated internationally as a vegetable. The fresh leaves, cooked or served raw in salads, are very tasty.

Odd as it may seem, New Zealand spinach was rarely eaten by Aborigines. According to anthropologists working north of Lake Eyre, its use there among Aborigines was "made known by the white man".

PIGFACES

Carpobrotus

OTHER NAMES: Karkalla (for *C. rossii*)

FIELD NOTES: Leaves are thick and fleshy, triangular in cross-section, often reddish. Flowers are purple with many shiny petals, and the fruits are purplish-red (rarely yellow) with two horns, ripening mainly in summer and autumn.

Pigfaces are common creepers of beaches, dunes and headlands. Eastern pigface (*C. glaucescens*) occurs in Queensland and New South Wales, karkalla (*C. rossii*) in the south, and western pigface (*C. virescens*) in Western Australia. Karkalla and inland pigface (*C. modestus*) also occur inland in South Australia and Victoria.

USES: Pigfaces are among Australia's tastiest wild fruits, having a soft wet pulp that tastes like salty strawberries or fresh figs. Aborigines thought highly of this fruit, as C. Wilhelmi recorded in 1860:

> Pressing the fruit between their fingers, they drop the luscious juice into the mouth. During the karkalla season, which lasts from January until the end of summer, the natives lead a comparatively easy life; they are free from any anxiety of hunger.

He noted that some tribes "as a substitute for salt with their meat, eat also the leaves of this saline plant". Settlers were also very fond of the fruit.

Sarcozona (*Sarcozona praecox*), a very similar plant of inland saline plains with edible fruit, is distinguished by the minutely warty surface of its leaves.

Round-leaved pigface or rounded noon-flower (*Disphyma crassifolium*) of southern beaches and plains, lacks succulent fruits, has purple petals with white bases, and less angular leaves, which are edible.

GOATS-FOOT CONVOLVULUS

Ipomoea pes-caprae subsp. *brasiliensis*

OTHER NAMES: Purple beach convolvulus, coast morning glory, *I. brasiliensis*

FIELD NOTES: This common creeper of beaches and dunes has thick folded leaves shaped like cloven hooves (hence *pes-caprae* meaning foot of goat), and pale pink or purple trumpet flowers. Because its seeds are spread by ocean currents, the plant has a world-wide distribution in the tropics. In Australia it occurs south to Sydney.

USES: Goats-foot convolvulus is closely related to sweet potato, and it too has a starchy edible root, though this is fibrous and unpalatable, sometimes stinging the throat. It was cooked and eaten by Aborigines and Pacific Islanders only as a standby food.

Young leaves were cooked and eaten in India and Japan, and make a pleasant vegetable for beachside campers. Aborigines did not use them.

Goats-foot convolvulus was an important bush medicine in northern Australia, Torres Strait, and throughout South East Asia. Heated leaves were placed on sores, stings and boils.

Mile-a-minute (*Ipomoea cairica*) is a closely related vine in Eastern Australia with leaves divided like a hand into five or seven slender segments (see leaf gallery), and purple, pink, or white funnel flowers. It grows mainly behind beaches and along creeks and gullies and is usually regarded in Australia as an introduced weed, although it is probably a pre-European introduction, at least in southern Queensland, where Aborigines were recorded eating the tubers.

Beach Bean

Canavalia rosea

Other Names: Fire bean, sea bean, wild jack bean, Mackenzie bean, [*C. maritima*]

Field Notes: Beach bean is a vigorous creeper with rounded leaves in threes, large bean pods, and pink or purple (rarely blue), pea flowers. Pods are 8–14 cm long, greenish and flattened at first, becoming brown and bulging, then splitting violently to catapult the hard seeds.

Beach bean grows on beaches, dunes and nearby open areas, south to about Kiama. It also occurs in Asia.

Uses: The first Englishman to eat beach beans was Captain Cook, who cooked up the beans during his enforced stay at Endeavour River in 1770. He said they were "not to be dispised" though Joseph Banks dismissed them as "a kind of beans, very bad". The beans were next eaten by Governor Phillip and his Surgeon-General John White, in 1788. White recorded that the beans "were well tasted, and very similar to the English long-pod bean". But the men evidently tasted the raw beans, for they were soon "seized with a violent vomiting".

The younger flat pods make a pleasant cooked vegetable, though they are inclined to be fibrous. The larger pods (illustrated) contain big green seeds, which Malaysians boil as porridge. After roasting in the pod, the big seeds are very tasty, resembling broad beans. Surprisingly, there are very few records of Aborigines eating the seeds. Most tribes in northern Australia today regard the plant as poisonous, which, in its raw state, it is.

T. implexicoma, *with large pale leaves, and* R. candolleana

Coastal Saltbush

Rhagodia candolleana

Other Names: Seaberry salt-bush, [*R. baccata*]

Field Notes: This is a shrub or tangled climber with fleshy leaves, mostly 1-2.5 cm long, of variable shape. The clustered, button-shaped fruits are about 5 mm wide and dark red to almost black when ripe. Coastal saltbush grows beneath trees or in hollows behind sandy beaches, and occasionally in woodlands further inland.

Uses: Aborigines reportedly ate *Rhagodia* berries in Victoria, including possibly this species, though the berries are very bitter. The cooked leaves are tender and succulent, and were probably among those eaten by early settlers.

Bower Spinach

Tetragonia implexicoma

Field Notes: This unmistakable creeper has broad, thick, pale green leaves, 2-6 cm long, which glisten, especially on their undersides. Small, long-stalked flowers with four yellow petals are followed by juicy, glistening, reddish fruits about 8 mm wide.

Bower spinach forms thick mats on bare sand behind beaches and cliffs, or hangs in curtains from seashore shrubs and trees. It is closely related to New Zealand spinach.

Uses: Settlers cooked up the succulent leaves like spinach, and Colonial botanist Baron von Mueller suggested the plant be cultivated as a vegetable. The salty-sweet fruits are edible.

MUNTARI

Kunzea pomifera

OTHER NAMES: Muntries, native apple, cranberry, crab apple, muntaberry.

FIELD NOTES: Muntari is a woody creeper of coastal dunes, found also in clearings in mallee and desert woodlands. The stiff, rounded leaves are 5–8 mm long, with sharp curled tips. The white flowers have many long stamens and small, rounded petals, like eucalypt blossoms. The purplish berries are 8–12 mm wide, furry-skinned, contain several tiny seeds and are produced in tight clusters which ripen in autumn.

USES: Delicious apple-flavoured muntari berries were probably a staple food of Aborigines between Yorke Peninsula and Portland Bay. Around the Glenelg River, Aborigines came from afar during the muntari season, as settler James Dawson recorded in 1881:

> In collecting the berries they pull up the plants, which run along the surface of the sand in great lengths, and carry them off on their backs to their camps to pick off the fruit at their leisure. On the first settlement of the district by sheep owners these berries were gathered by the white people, and they made excellent jams and tarts.

Further east, in the South Australian Coorong, Aborigines pounded and dried excess fruits into large cakes that were stored for eating during winter, when other foods were scarce, or were traded with other tribes for stones for axe-making.

In the Millicent district the fruits are still gathered and made into jams which are sold at country fetes. Muntari has been introduced into horticulture and may be bought from nurseries to use as a decorative ground cover.

MIDYIM

Austromyrtus dulcis

OTHER NAMES: Midgen berry, silky myrtle

FIELD NOTES: Midyim is a small 1-2 m shrub with slender opposite leaves, easily identified by its characteristic berries. The leaves are 1-3 cm long, and young foliage is coppery-coloured. Small white flowers are followed by soft, speckled white berries, about 1 cm wide, containing small seeds, ripening in summer and autumn.

Midyim is a common understorey shrub of heathlands and woodlands, growing in sand by the sea, from Grafton to Fraser Island.

USES: Midyim is one of Australia's tastiest wild fruits, with a soft, sweet, slightly aromatic pulp that melts in the mouth. It was a favourite of Aborigines in Moreton Bay, as described by Thomas Welsby on Stradbroke Island early this century:

> It springs up and grows like a wheat field . . . one can go through acres of the shrub with its white, sweet-tasting berry until stopped by lagoon or salt water. It is the most sought-for berry or fruit on the island. Children will collect it by the tin-full, and even the elders will join with gusto in its eating.

Quaker missionary James Backhouse, a colonial wild foods' buff, described the berries as "the most agreeable, native fruit, I have tasted in Australia; they are produced so abundantly, as to afford an important article of food, to the Aborigines".

Midyim has become popular in horticulture, for it is hardy and adaptable and the fruits and foliage are decorative. It is well worth growing for its tasty fruits.

The related scaly myrtle (*A. hillii*), a rainforest tree, has black edible fruits.

Grey Saltbush

Atriplex cinerea

Other Names: Coast saltbush, barilla

Field Notes: Grey saltbush is a shrub of beaches and saltmarshes, readily recognised from afar by its pale, grey or blue-grey foliage. The leaves are blunt-tipped, 2.5–4 cm long. Male plants produce lumpy brown spikes on the tips of branches; female plants (illustrated) produce clusters of small grey pods 6–10 mm long.

Grey saltbushes grow scattered along foreshores, and are always closer to the sea than other shrubs. They form squat shrubs 1–2 m tall.

Uses: Grey saltbush was an important vegetable of Australian colonists, first eaten by the hungry officers of the First Fleet, one of whom described it in his journal as "a sort of sage". During a famine in Tasmania in 1807, when kangaroo flesh sold for one shilling sixpence a pound, the leaves, known as "Botany Bay Greens" were "the chief support of the inhabitants", according to colonial botanist Daniel Bunce. Historian Daniel Mann wrote in 1811 that Botany Bay greens were "esteemed a very good dish by the Europeans, but despised by the natives". In Tasmania, and on the islands of Bass Strait, vast acreages of this plant were cut and burnt to make soda for soap-making. The boiled leaves have a very tender, succulent taste.

There are many related species of saltbush, found either on beaches or inland saline plains, with succulent, edible leaves. These should always be cooked before eating.

COAST BEARD HEATH

Leucopogon parviflorus

OTHER NAME: White currants

FIELD NOTES: This is a shrub, rarely a small tree, with creamy fruits about 5 mm wide, clustered along stalks about 1 cm long. The slender leaves are 1-3 cm long, with recurved margins, and are green on both sides. The tiny white flowers have five furry petals.

In southern Australia coast beard heath is very common on dunes, headlands and cliffs by the sea, often forming dense thickets (at Port Campbell, for example). It springs up on road verges and other sites of human disturbance, growing up to several kilometres inland in heaths, mallees, and along road cuttings in forestry reserves.

In eastern Australia it grows as small scattered shrubs on coastal dunes, as far north as Fraser Island.

Coast beard heath is a member of the large heath family (Epacridacae), described in detail on page 128. It closely resembles the prickly broom heath.

USES: The tiny fruits have a pleasant lemony taste, and make a refreshing snack food. They are relished by seagulls which, too heavy to perch on the shrubs, are forced to half hover, half perch, in flocks of up to 80 birds—a most ungainly spectacle. The gulls help spread the seeds. Further inland, silver-eyes and blue wrens disperse the seeds.

BOOBIALLA

Myoporum

OTHER NAMES: Boobyalla, water bush, native juniper, native mangrove, blueberry

FIELD NOTES: Boobiallas are very variable shrubs or trees with small, white, five-petal flowers, smooth alternate leaves and shiny purplish fruits, 5–8 mm wide, containing one stone. There are four coastal species.

Southern boobialla (*M. insulare*, illustrated) is a stout, rough-barked tree and is the largest tree found on southern shores, though the twisted trunk often trails along the ground. Its white flowers are purple-spotted, the fruits are bluish-purple, and the leaves are usually serrated near the tips.

There are three eastern species, all with reddish-purple fruits, often sparsely produced. One of these, western boobialla (*M. montanum*) is shown on page 171. The similar *M. acuminatum*, a small tree found on southern Queensland and New South Wales coasts, differs by having spotted flowers. *M. boninense* (previously called *M. ellipticum*) is a squat, often sprawling shrub, with relatively fleshy, shiny, sometimes very broad leaves, and unspotted flowers. It grows along the Queensland and New South Wales coasts, often on exposed headlands.

Boobiallas grow on dunes, coastal headlands, estuaries and mangrove fringes. Identification of the different species has been very confused in the past.

USES: Colonial botanist Joseph Maiden drew attention to the edible fruits, which are inclined to taste bitterly aromatic and salty sweet. Some crops are more palatable than others. A food of birds, they are perhaps best left to the birds.

YELLOW PLUM

Ximenia americana

OTHER NAMES: Wild apricot, sea lemon

FIELD NOTES: Yellow plum is an untidy shrub or small sprawling tree bearing plum-like fruits. Leaves are oval in outline, bright green on both sides, and smell of bitter almonds when crushed. The stems often carry small spines. Pale flowers with four arching furry petals are followed by shiny, lemon-yellow or orange-red fruits about 3 cm long. These consist of a thin layer of succulent flesh surrounding a large, smooth, pale brown seed.

Sea lemon grows amongst vegetation on and behind beaches, and also occurs in central Queensland in woodlands and vine scrubs.

USES: The fruits have a pleasant plum-like flavour. In Asia, where this shrub also grows, the young leaves have been cooked as a vegetable and the oily seed kernels have been eaten in small amounts. They contain cyanide and should not be taken in quantity.

There are two other, similar-looking edible fruits found on small trees growing on tropical beaches. The tanjong (*Mimusops elengi*), found from north-western Australia to north Queensland, has darker leaves, and orange-red, long-stalked fruits, 1.3-1.5 cm long, containing several shiny, orange-red, wedge-shaped seeds.

The wongi (*Manilkara kaukii*), found only in north Queensland, has very tasty, orange-red, 3-4 cm long fruits, containing several long, shiny seeds. The leaves are rigid, blunt-tipped, dark-green above, pale and silky below. Wongi plums are popular among Torres Strait Islanders, who travel from island to island to harvest the crop. North Queensland Aborigines used to bury the fruits to hasten ripening for big social gatherings.

GREAT MORINDA

Morinda citrifolia

OTHER NAMES: Cheese fruit, Indian mulberry

FIELD NOTES: Great morinda is a shrub or small tree, found scattered along beaches and coastal headlands, and in monsoon rainforest along lowland streams, where the fallen fruits can be found on the forest floor. The bright green leaves are large (10-30 cm long), shiny, opposite, and heavily veined. Heads of tubular white flowers are followed by grotesque warty fruits, 4-10 cm long. Ripe fruits are very soft and smelly, with a translucent white skin. They may ripen at any time of year.

Great morinda seeds are spread by ocean currents, and the tree is found on beaches as far away as India, Tahiti and Hawaii.

USES: Great morinda fruits produce one of the most disgusting smells in the Australian bush, comparable to rotting cheese. The fruit is nonetheless edible, with a flavour combining camembert cheese and custard apple. The mushy pulp is a good source of vitamin C (see Chapter 10).

Once an important fruit of Aborigines, great morinda is less popular today. It is sometimes eaten slightly unripe, while still crisp and odourless. In the Northern Territory it enjoys a strong reputation as a cure for colds; the fruit is simply eaten raw.

In Asia and New Guinea, great morinda leaves have been cooked and served as a vegetable; but Aborigines did not eat them. The roots yield a yellow pigment, a traditional batik dye, and trees were cultivated in Asia for this purpose. Aborigines fixed this dye in seawater for colouring their bags and aprons.

COAST WATTLE

Acacia sophorae

OTHER NAMES: *A. longifolia* var. *sophorae*

FIELD NOTES: The leaves are broad, (not sickle-shaped), blunt-tipped, 5-12 cm long, 1.2-4 cm wide, with two to five longitudinal veins. They are held rigidly out from the stems and do not dangle like many wattles. Fluffy yellow flower spikes in winter and spring are followed in summer by curly, green, cylindrical, bean-like pods, 5-15 cm long, which turn brown when ripe.

Coast wattle is the common wattle on beaches in south-eastern Australia. It is a large sprawling shrub found behind beaches, in dunes, and in woodlands on sandy soil.

USES: Quaker missionary James Backhouse observed in 1843: "The natives of Tasmania used to roast the ripening pods of this wattle, pick out the seeds and eat them." C. W. Schurmann in 1879 described how the nondo bean (coast wattle) was so valued that "the Kukata tribe, notorious for ferocity and witchcraft, often threaten to burn or otherwise destroy the nondo bushes in order to aggravate their adversaries". Elderly Aborigines alive today from Wreck Bay, south of Sydney, remember their elders eating the steamed seeds—they taste rather like green peas. The pods are irritant and inedible.

The Wreck Bay Aborigines also remember the eating of steamed golden wattle (*A. longifolia*) and sweet wattle (*Acacia suaveolens*) seeds. The seeds of other wattles were very popular Aboriginal foods in the outback.

P. tectorius

Pandanus

Pandanus

Other Names: Screwpine, breadfruit, wynnum

Field Notes: Pandanus "palms" are distinctive trees with very long, leathery, serrated-edged leaves, and huge globular fruit structures.

The coast pandanus (*P. tectorius*, formerly *P. peduncularis*) has stilt roots (see illustration) and grows by the sea in eastern Australia. The northern pandanus (*P. spiralis*) has a spirally patterned trunk, no stilt roots, and grows in woodlands and by the sea in northern Australia. The Kakadu pandanus (*P. basedowii*) has stilt roots and grows only on the rocky scarps of western Arnhem Land.

Identification of pandanuses is confused, and there are known to be other edible species, including one in north Queensland rainforests with sticky red edible pulp.

Uses: The large pandanus fruit is made up of many hard orange wedges, 5-10 cm long, which topple to the ground when ripe. Each woody wedge contains a few slender seeds which taste deliciously nutty, raw or cooked. Though difficult to extract, the seeds were an important Aboriginal food, for they are very rich in fat (44-50 per cent) and protein (20-34 per cent).

Anthropologist Betty Meehan found that Arnhem Land Aborigines could process enough seeds in two hours to feed someone for more than a day. Pandanus wedges were split by axe and the seeds prised out with a djin-garn-girra, a unique steel-tipped tool that may date back 300 years to early Indonesian trade. Aborigines elsewhere in the north prised out the seeds with slivers of wood, bone, or stingray spines, first roasting the wedges to make them brittle, then splitting them with stone axes. The process is tedious without steel tools.

The wedges of both the coast and northern pandanuses have a fleshy base imbued with sweet-smelling orange pulp. The strong sweet flavour compares with sweetened baked potato. But as explorer Ludwig Leichhardt discovered, it should not be eaten raw:

> I had frequently tasted the fine looking fruit of the pandanus but was every time severely punished with sore lips and

P. tectorius *fallen wedges*

> blistered tongue; and the first time that I ate it, I was attacked by violent diarrhoea. I could not make out how the natives neutralised the noxious properties of the fruit; which formed no small part of their diet.

Aborigines baked the fruits in hot sand or ashes to remove the irritant, but were sometimes able to eat them raw, perhaps by developing some immunity. Along the Roper River the sweet pulp was beaten from the fruits and soaked for a few days to make a mild alcoholic drink.

The raw inner bases of pandanus leaves can be nibbled as a snack. The tough fibrous leaves were torn into strips and woven into baskets, mats, arm bands and ropes, and the logs were roped together to make rafts.

P. tectorius

SEA ALMOND
Terminalia catappa

OTHER NAMES: Indian almond, tropical almond, northern bush almond.

FIELD NOTES: Sea almond is a stately tree of tropical beaches with very large leaves and big purple fruits. The leaves are oval, bright green above, paler below, 12-36 cm long, 6-15 cm wide and have prominent veins. They turn red before being shed. Sprays of tiny white flowers are followed by fruits about 5-8 cm long, with a ridge around the widest edge. They ripen mainly from February to May.

Sea almond is found on sandy and stony beaches north of Bowen. It is often planted in parks in the tropics.

USES: The sea almond fruit contains a large woody stone which can be cracked (with difficulty) to extract a tiny white kernel; this is rich in protein, oil and thiamine, and has a strong almond taste. It is a popular snack food of Aborigines and Torres Strait Islanders. The sea almond has no doubt been spread deliberately throughout Torres Strait, for the trees can be seen growing on beaches beside most villages. Large rocks for cracking the stones are kept beneath the trees. The purple pulp of the fruit is also edible and has an unusual tropical fruit flavour.

The related nutwood (*T. arostrata*) and *T. grandiflora*, two small-leaved trees found in woodlands in north-western Australia, have edible kernels.

GREY MANGROVE

Avicennia marina

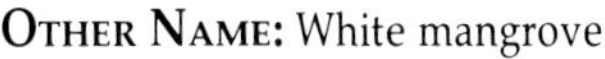

OTHER NAME: White mangrove

FIELD NOTES: Grey mangrove is the common mangrove of estuaries and sheltered coasts, easily recognised by its many black, spike-like aerial roots, which sprout from the mud near the tree and draw in essential oxygen absent from the sodden mud below. Leaves are 4-12 cm long, opposite, broad or slender, with a greyish, yellowish or silvery underside. The "seed pods" are 1.5-4 cm long.

USES: A tidal mudflat is a hazardous place for a germinating seed—salty, infertile, and very unstable. Mangroves solve the problem by producing large, specialised seed structures, packed with food reserves, resistant to salt, sea and sun. Northern Aborigines learned to harvest these starchy foods, first leaching out distasteful tannins and bitter substances which are the plant's chemical defence.

Several methods of preparation were known. Moistened seeds could be heaped into stone ovens, sealed with bark, baked for two hours then soaked in a pool dug in sand before eating. Alternatively a "sweet" pulp was prepared by baking, pounding and sifting the pods through dillies. Although bland, these seeds were an important wet season food.

Grey mangrove is found south to Melbourne, but its seeds were used as food only in the tropics. The cigar-shaped pods of another mangrove, the orange mangrove (*Bruguiera gymnorhiza*), were also a staple food in north Queensland.

Queensland University's Botany Club has promoted grey mangrove pods as a gourmet item. "Avicennia fruit dip"made from the pods boiled in three changes of water, tastes revolting.

Nardoo

Marsilea drummondii

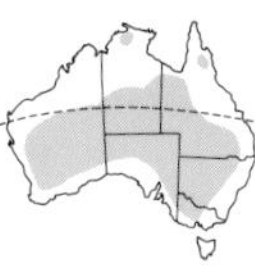

Other Names: Southern cross, clover fern

Field Notes: Nardoo is a small fern resembling four-leaf clover. The four leaflets are hairy or hairless, on stalks 2-30 cm tall; the spore capsules (sporocarps) are furry pods, 5-9 mm long, with two small knobs at the back, on stalks 1-6 cm tall.

Nardoo grows in colonies on the edges of claypans, swamps, river flats and ditches. When growing in water the leaves float, but on dry land are carried erect. Only plants on dry land produce sporocarps, which lie dormant until floods come, when they split to release the spores.

Uses: Nardoo is infamous as the food on which Burke and Wills starved to death. "Starvation on nardoo is by no means very unpleasant, but for the weakness one feels and the utter inability to move oneself, for, as far as the appetite is concerned, it gives me the greatest satisfaction", Wills wrote in 1861, shortly before his death.

Aborigines ground nardoo sporocarps between stones, and on removing the husks were left with a yellow spore flour that was moistened and baked. Nardoo cakes were said to be astringent, indigestible and low in nutrition, though surprisingly, their nutritional content has never been assessed.

Records of nardoo use come from the catchments of the Cooper, Diamantina and Darling Rivers, where Aborigines thrived on wild grains, pigweed, and to a lesser extent, nardoo. Elsewhere in Australia it was considered to be inedible.

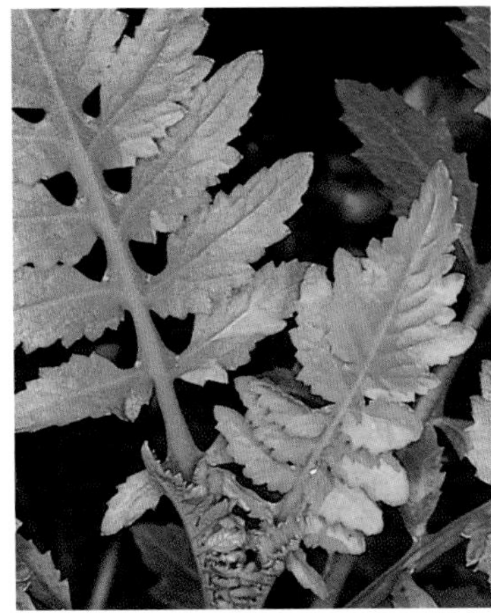

MARSH CRESS

Rorippa palustris

OTHER NAMES: Yellow cress, [*R. islandica*]

FIELD NOTES: Marsh cress produces clusters of tiny flowers with four yellow petals, atop stalks carrying tiny sausage-shaped pods, 4–13 mm long. The soft leaves are 3–12 cm long, deeply lobed, with serrated edges. The stems grow 30–100 cm tall, often becoming reddish and wiry with age.

Marsh cress is an untidy herb of muddy river and stream margins, ditches, and boggy paddocks. It often grows beside drains in low-lying suburbs of cities and towns.

USES: Several native cresses are found in south-eastern Australia, and bush pioneers gathered these as food. Colonial botanist Joseph Hooker stated that marsh cress "and other species afford excellent pot-herbs when luxuriant and flaccid". In Tasmania this cress was said to form "a good salad"and along the Nepean River it was known as "native cabbage". Aborigines in Victoria, and probably elsewhere, ate the leaves, which have a slightly spicy taste.

Marsh cress is a native plant in Europe and Asia, and many botanists believe it to be an introduced weed in Australia. But Joseph Banks collected the species in New Zealand in 1769, suggesting that the seeds probably came to this region long ago on the feet of migrating waders.

The other cresses eaten by colonists (*R. dictyosperma, Cardamine gunnii, C. tenuifolia*) have tiny white flowers with four petals, cylindrical seed pods, and slender delicate leaves of varying shape. They grow in south-eastern Australia in damp forests or near streams.

Kang Kong

Ipomoea aquatica

Other Name: Water spinach

Field Notes: The showy flowers are pink, purple or white with a dark purple centre. The leaves are slender, 4.5–12 cm long, heart-shaped at the base, tapering at the tips, and are on long stalks.

Kang kong is a trailing vine found around the edges of swamps, billabongs and dams, either floating in the water or creeping over nearby mud or wet sand. It also grows on boggy floodplains among grasses and sedges. It is native both to Asia and northern Australia.

Uses: In South East Asia kang kong is a very popular vegetable, widely cultivated in ponds and in even the most polluted of inner-city canals, where it forms thick mats of lush floating foliage. The young leaves and shoots are sold in bundles in markets as a soup vegetable. Rich in vitamin C, they yield up to 100 milligrams of ascorbic acid per 100 grams, twice that of oranges. Aborigines did not eat many kinds of leaves and made no use of kang kong.

Kang kong is closely related to sweet potato (*I. batatas*), and the latter also has edible leaves—a popular village vegetable in Asia. There are many closely related vines found in open forests in northern Australia, recognised by their similar trumpet flowers, which produce edible tubers.

WATER RIBBONS

Triglochin

OTHER NAMES: Swamp arrowgrass, creek lily

FIELD NOTES: These water plants have very long, slender leaves that vary greatly. Some forms have limp leaves that float on the water, as illustrated, or arch slightly above the water, while others have stiff leaves that rise upright from the water. The leaves are usually 2-40 mm wide (sometimes much wider) and about 30-225 cm long. The best guides to identity are the tall, greenish, cylindrical seed spikes, and the rounded or cylindrical tubers, 0.5-14 cm long, 0.5-1.5 cm wide, on the tips of some roots.

In the past, all tuberous water ribbons were classified as *T. procerum* (or *T. procera*), but eight species are now recognised. It is likely that all are edible, although only *T. procerum* of south-eastern Australia and *T. dubium* of northern and eastern Australia are recorded as Aboriginal foods.

Water ribbons grow in most freshwater situations, including muddy outback billabongs, soaks behind beaches, seasonally flooded hollows, sluggish rivers, swamps, and fast-flowing streams. They flourish in water up to 2 m deep, but also sprout in wet mud and on sandbanks.

USES: Aborigines had their own baby foods, and water ribbons was one of these. On Groote Eylandt the bland starchy tubers were roasted, pounded and fed to teething babies and the elderly. The raw or roasted tubers were also eaten by adults, and were probably an important staple food throughout much of Australia. Up to 200 tubers are borne by a single plant.

RUSHES AND SEDGES

family Cyperaceae

Looks can be deceiving: the sombre sedges illustrated here were among the most important of Aboriginal foods, producing underground stores of starch which Aborigines baked and ate. Sedges may grow in vast colonies and their tubers, though small, can be gathered in large numbers. Some were staple foods. They are all edible raw, and cannot be confused with any poisonous plant.

Although nondescript, the edible rushes and sedges are surprisingly easy to idenify. Club-rushes (*Bolboschoenus*) and nalgoo have grass-like leaves but the others have leaves reduced to minute scales, easily overlooked, and their cylindrical stems take the place of leaves. These are large sedges, a metre or two tall, but some sedges are small, especially nalgoo and sea club-rush.

Spike rush and nalgoo were staple foods, and are still eaten in northern and central Australia respectively. The other sedges grow in the south and east, from where Aborigines were long ago exterminated, and their importance in Aboriginal diet is difficult to assess, for the old references to "rushes" are often ambiguous. Grey sedge and river club-rush have woody underground parts and were perhaps only famine foods, but the club-rushes are tasty, and were probably staple foods.

Some sedges produce starchy seeds, and tribes ate these after grinding them to flour. The stems were woven into baskets. Some of these sedges grow overseas and were used elsewhere: nalgoo and edible spike rush are eaten in Asia, and the Chinese cultivate grey sedge for weaving. All the tuberous sedges in Australia known to be edible are featured here.

GREY SEDGE

Lepironia articulata

FIELD NOTES: The rigid grey stems are 0.6–2.3 m tall and the single scaly seed spike is 1–3 cm long, produced near the stem tip. This very large, stiff sedge can be found lining the edges of wallum lakes and sluggish streams in low-lying coastal districts, and occasionally inland. It can be recognised from afar by its size and greyish colour.

USES: North Queensland Aborigines reportedly harvested the woody underground stems.

RIVER CLUB-RUSHES

Schoenoplectus litoralis, S. validus

OTHER NAME: [*Scirpus*]

FIELD NOTES: These rushes are distinguished by their soft (not rigid or wiry), easily flattened, greyish hollow stems, 0.6–1.6 m tall, and clusters of brown seed spikes, each 0.5–1.5 cm long. River or lake club-rush (*S. validus*), grows on edges of rivers, lakes and ditches. The very similar littoral club-rush (*S. litoralis*) grows along coastal rivers and creeks, usually in brackish water.

USES: North Queensland Aborigines ate littoral club-rushes after roasting and hammering the underground stems. Of river club-rushes we have no record, though American Indians harvested one variety.

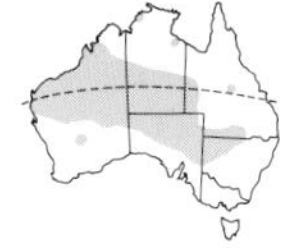

NALGOO

Cyperus bulbosus

OTHER NAMES: nutgrass, bush onion [misidentified previously as *C. rotundus*]

FIELD NOTES: Nalgoo closely resembles the garden weed, nutgrass, but bears shiny spindle-shaped tubers, 1–1.5 cm long (see tuber gallery) at the tips of its thread-like roots. The triangular stems are up to 40 cm tall, leaves are grass-like, and the seed spikes are reddish-brown. Nalgoo grows in colonies in mud or sand beside lakes, claypans, rivers and streams. The leaves wither as water levels recede, leaving only the buried tubers.

USES: Nalgoo was a staple of outback Aborigines. The pleasantly starchy tubers ("nuts") were dug with the hands or with short sticks, and eaten raw or roasted.

B. fluviatilis

B. caldwellii

CLUB-RUSHES
Bolboschoenus

OTHER NAMES: Bulrushes, [*Scirpus*]

FIELD NOTES: Leaves are shiny and grass-like, and the stems are triangular in cross-section. Each stem is topped by a cluster of scaly brown seed spikes, 1-2 cm long, and a few grass-like "leaves" (bracts).

There are three species. Sea club-rush (*B. caldwellii* formerly *Scirpus maritimus*) is the smallest, growing only 0.45-1 m tall, and producing one to ten seed spikes in a tight cluster. *B. medianus*, and especially the marsh club-rush, or river bulrush (*B. fluviatilis*) are much larger, and their numerous seed spikes are borne on long stalks in sprays of up to 50 spikes. Marsh club-rush grows up to 2 m tall and has leaves up to 17 mm wide.

Club-rushes sprout in colonies on the edges of sluggish rivers, swamps and ditches. Sea club-rush also grows on saltmarshes, tidal streams, and low-lying waste ground.

The underground stems of club-rushes creep through mud, swelling now and then into rounded tubers from which clusters of new leaves and roots sprout (see tuber gallery). Tubers are at first white and starchy with a sweet, coconut milk flavour, but soon become black and woody.

USES: Explorer E. J. Eyre told of Aborigines along the Murray River eating the walnut sized tubers of "belilah"—probably marsh club-rush. The tubers were roasted, pounded and made into cakes.

Wet (late autumn). The plants die away in winter, surviving the Dry as tubers.

Spike rush flourishes in ditches and seasonal swamps in low-lying areas. It is common at Kakadu and around Townsville.

USES: The tubers or "nuts" were staple Aboriginal foods, dug by women during the Dry (mainly winter and spring). On Mornington Island the best rush areas were personally owned, and stems were knotted to show this. The husked tubers are crisp, juicy and sweet and are edible raw or roasted. They were sometimes beaten into cakes that could be stored for up to a fortnight. The water chestnuts used in Chinese cooking come from the same plant.

EDIBLE SPIKE RUSH

Eleocharis dulcis

OTHER NAMES: Chinese water chestnut, bush nut, bulguru

FIELD NOTES: Stems are dark green, hollow, tubular, partitioned inside, thin and pliable (not stiff), of pencil thickness, 1-2 m tall, forming large tangled clumps. The scaly seed spike sprouts atop the stems, with scales 6-6.5 mm long. Small onion-shaped tubers are produced at the end of the

TALL SPIKE RUSH

Eleocharis sphacelata

FIELD NOTES: (Not illustrated.) This rush resembles edible spike rush but the seed spike scales are larger, 8-8.5 mm long, and the plant has no tubers but a starchy trailing underground stem to which the hollow green stems are attached.

USES: Aborigines ate the starch in the young underground stems.

BULRUSH

Typha

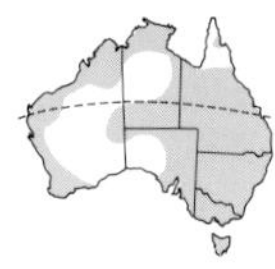

OTHER NAMES: Cumbungi, wonga, reed-mace

FIELD NOTES: The strap-like leaves are 1–2 m tall and the seeding spike looks like a sausage on a skewer. Australia has two difficult to distinguish species: *T. orientalis*, the common bulrush in the south, and *T. domingensis*, more common in the north.

Bulrushes flourish in sluggish streams, swamps, lakes, billabongs, wet hollows, dams, and irrigation channels.

USES: Bulrush starch was once a staple food of Aborigines, especially along the Murray River and its tributaries. On the Lachlan River, explorer Major Mitchell saw "natives" carrying on their heads great quantities of the "roots" (actually underground stems): "They take up the root of the bulrush in lengths of about eight or ten inches," he wrote in 1839, "peel off the outer rind, and lay it a little before the fire, then they twist and loosen the fibres, when a quantity of gluten exactly resembling wheaten flour, may be shaken out." The starch can also be eaten raw, and the leftover fibres were often spun into tough string.

Some tribes ate the raw green young flowerstalk during spring. Settler P. Beveridge wrote: "The natives are extremely partial to it, they therefore consume it in great quantities."

Aborigines apparently made no use of bulrush pollen, though in India and New Zealand this was baked or steamed into nutritious cakes. All but one of the observations on Aboriginal use of bulrush are from temperate Australia, within the range of *T. orientalis*, suggesting it may have been the main species used.

TARO
Colocasia esculenta

FIELD NOTES: The enormous heart-shaped leaves are up to 40 cm long, carried on thick fleshy stalks up to 1 m long. The spikes of tiny greenish flowers are framed by orange or yellow spathes.

Taro differs from cunjevoi (*Alocasia macrorrhizos*, see page 84) in having its leaf stalk attached, not to the upper edge of the leaf, but to the back of the leaf about a third of the way from the top. Taro also has greenish, not red berries.

Taro grows in colonies beside lowland streams and lagoons, usually beneath a jungle canopy. In eastern Australia, as far south as Sydney, it occasionally runs wild as a garden escape. Feral plants are of overseas origin and often have purplish stems. They can be confused with other garden escapes, such as blue taro (*Xanthosma violaceum*), which has violet stalks and leaf veins.

USES: Taro may be one of those food plants that early seafarers introduced to Australia hundreds or thousands of years ago. A staple food of Polynesians and Melanesians, it is cultivated throughout the Pacific region, and in parts of Asia. Australian Aborigines did not farm taro, but cooked and ate the starchy underground stems of the wild plants.

Captain Cook ate taro leaves and "Yamms" while camped at Endeavour River. On 29 June 1770 he reported. "The Tops we found made good greens, and eat exceedingly well when Boil'd, but the roots were so bad that few besides myself could eat them." The boiled leaves are occasionally eaten as a vegetable overseas, but leaves I tried in north Queensland stung my throat.

WATERLILY

Nymphaea

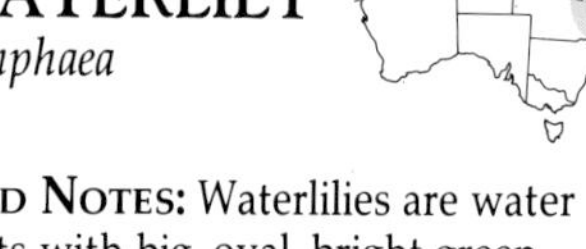

FIELD NOTES: Waterlilies are water plants with big, oval, bright green, floating leaves and large showy flowers with many petals. Flowers are blue, purplish, pink or white, depending upon the species.

Waterlilies grow in lagoons, swamps and sluggish rivers and streams. Australia has about seven native species, of which *N. gigantea* and *N. violacea* are the most widespread. The waterlilies growing wild around Brisbane and Sydney are introduced species.

USES: Waterlilies were a handy source of food for Aborigines. The tubers, seeds, flowers and stalks were eaten. Aboriginal women dived into lagoons to retrieve the golfball-sized tubers which sprout beneath the plant. These were roasted and eaten. Some kinds of tubers in north Queensland, with hairy roots on the surface, or yellow flesh, are "cheeky" (poisonous), and need to be leached in water before eating.

The tiny seeds are produced abundantly (up to 3,000 or more per pod) in apple-sized pods found ripening just below the water surface. Aborigines gathered the soft, under-ripe pods and roasted them before extracting and eating the delicious seeds. Young seeds extracted from the base of the flowers are also edible.

Aborigines drank nectar from waterlily flowers, though this can cause headaches if too much is drunk. The buds and peeled flower stalks, which taste like celery, were also eaten.

Marshworts (*Nymphoides*) resemble waterlilies, but are much smaller plants, producing flowers with only five, fringed, usually yellow, petals. Marshwort tubers have incorrectly been listed as Aboriginal foods, but marshworts do not produce tubers.

LOTUS
Nelumbo nucifera

OTHER NAMES: Pink waterlily, sacred lotus lily, nangram lily

FIELD NOTES: The sacred lotus has enormous rounded, dark green leaves, 20-90 cm wide, usually carried above the water, on stiff prickly stalks. The spectacular showy flowers are deep pink with a yellow centre, 15-25 cm wide, carried on prickly stalks up to 2 m tall. The seeds are about 2 cm long, carried in spongy, flat-topped pods.

The lotus grows in deep, permanent lagoons and billabongs. Its distribution is very patchy. The southernmost colony, at Nangram Lagoon near Condamine on the Darling Downs (where the photo was taken), has since died out. The lotus also grows very widely in Asia, as far north as Japan and the southern USSR. In India, China and Japan it is planted in temple gardens as a sacred plant. The seeds are extremely long-lived, with germination reported after 237 years.

USES: Aborigines ate the young roasted tubers, the raw or roasted seeds, and the inner leaf stalks. The women often had to dive deep under the water to retrieve the tubers, which are sweet-tasting but fibrous. The large seeds were sometimes ground into meal and baked as bread. The explorer Ludwig Leichhardt, by roasting and pounding the seeds, made "an excellent substitute for coffee". In Asia the seeds are eaten raw, boiled or roasted, and the young leaves are boiled or eaten raw as a vegetable. Australian florists sell the dried pods for flower arranging.

BUNGWALL FERN

Blechnum indicum

OTHER NAME: Swamp water fern

FIELD NOTES: Bungwall is a tall and coarse fern of low-lying coastal swamps, identified by its large size, serrated leaf margins, and long black-skinned rhizomes (underground stems), packed with slimy white starch. Ferns growing in water produce fronds 1-2 m tall and rhizomes 2-3 cm thick; ferns on land are smaller.

USES: From accounts of colonists it is obvious that bungwall rhizome was the staple plant food of Aborigines in Moreton Bay. In 1894 Dr Bancroft recorded that "it is first dug out with a sharpened stick, dried in the sun for a short time, roasted and afterwards bruised, when it is ready to be eaten in conjunction with fish, crabs and oysters". Stradbroke Island pioneer Thomas Welsby recalled a kind of Aboriginal "biscuit" prepared from the dried and powdered rhizomes. Constance Petrie wrote: "It was mostly the gins who dug this up and put it in their dillies to carry to camp; great loads there would be at times, for the root was highly esteemed." Special stones were used to soften the rhizomes, and the "chop-chop" of the stones was a familiar campsite sound.

Bungwall must have been an easily gathered staple when the long rhizomes were trailing freely in water, and a day's supply was probably gathered in an hour or so. The shipwrecked convicts Pamphlet, Finnigan and Parsons were able to subsist on the rhizomes while traversing Moreton Bay, and when staying with Aborigines were liberally supplied with the fern. The very bland starch can be eaten raw.

NATIVE LASIANDRA

Melastoma affine

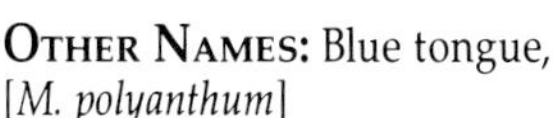

OTHER NAMES: Blue tongue, [*M. polyanthum*]

FIELD NOTES: The distinctive leaves are coarse and hairy with three prominent veins. Flowers are rosy pink or purple (rarely white); fruits are brown scaly capsules which burst when ripe to expose the purplish-black pulp. This is a shrub, to 2 m tall, of tea tree swamps, sandy stream margins near the sea, and in the north, of monsoon rainforest near creeks. (It also appears on the previous page, growing left of the bungwall.)

USES: Aborigines eat the sweet (sometimes bitter) pulp, which stains the mouth blue, hence "blue tongue".

MANGROVE FERN

Acrostichum aureum, A. speciosum

FIELD NOTES: (Not illustrated). These ferns resemble bungwall, but the leaflets are very large and leathery, with smooth (not serrated) edges. *A. aureum* has fronds up to 4 m tall and very blunt-tipped leaflets; it grows only in swamps and near mangroves in north Queensland (and overseas). *A. speciosum* has fronds up to 1.5 m tall, and more tapered leaflets. It grows along river estuaries and in marshes close to mangrove swamps, from New South Wales to Western Australia.

USES: The thick black underground stems contain a small store of starch which was eaten by Aborigines. The curled shoots (called croziers or fiddles) can also be cooked and eaten.

LEICHHARDT TREE

Nauclea orientalis

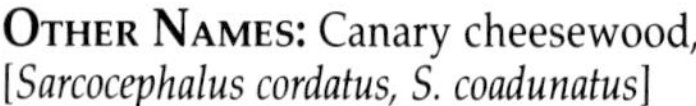

OTHER NAMES: Canary cheesewood, [*Sarcocephalus cordatus, S. coadunatus*]

FIELD NOTES: This is a majestic tree with big, dark green leaves, horizontal tiered branches, and smelly brown fruits. The leaves are opposite, oval, velvety to touch, mostly 20-30 cm long, with distinctive cross-veins. Yellow spherical flowerheads are followed by coarse, yellowish-brown fruits, 3-5 cm wide, containing many tiny seeds, ripening from February to May.

Leichhardt tree grows in lowland areas along streams and swamp margins, rainforest fringes, and in jungles behind beaches.

It is very conspicuous along creeks because of its large leaves.

USES: The soft, smelly fruits of the Leichhardt tree are alluring to bats, which eat the fruits and help disperse the seeds. The horizontal layered branches allow access to clumsy fruit bats; the strong smell of the fruits makes them easy to find.

The mushy, bitter, banana-like pulp was once a food of Aborigines. It can be high in vitamin C, but researchers have found surprising variations—from as much as 29 milligrams per 100 grams to a mere 1.4 milligrams in only three samples tested.

The bitter bark of this tree was used by Aborigines to poison and catch fish, and to induce vomiting following snakebite (a useless treatment). Bushmen compared the bitterness with quinine, and prescribed the bark to treat malaria, though this also was useless. Arnhem Land Aborigines made canoes from the trunks.

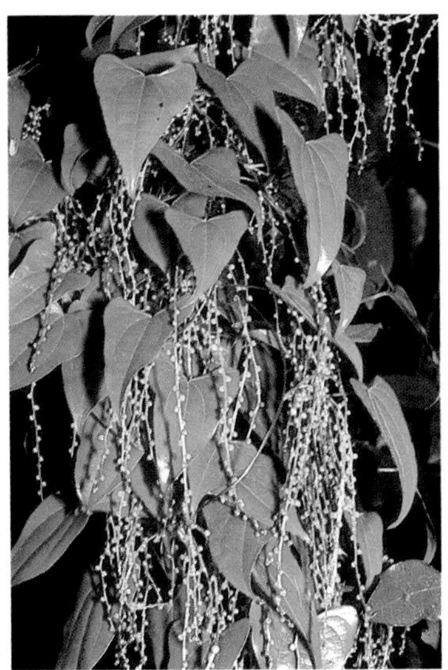

RAINFOREST LONG YAM

Dioscorea transversa

OTHER NAME: Pencil yam

FIELD NOTES: Long yam is a vine with very thin, wiry, coiling stems and shiny leaves with five prominent veins, 5-13 cm long, which are more or less heart-shaped (see leaf gallery for variation). The minute buds (illustrated) and flowers hang in strings; they are followed by clusters of hanging pods, 2-3.5 cm long, which have three rounded wings. The similar round yam (*D. bulbifera*) has narrower pods and different leaf venation.

Long yam grows around rainforest edges and clearings as far south as Stanwell Park, just south of Sydney. It also grows in shady gullies in more open forests. It is common but easily overlooked. In northern Australia it grows in open forests in sandy soils, often behind beaches. This northern form has less shiny leaves and produces bulbils, and is probably a separate, unrecognised species. It is illustrated on page 121.

In north Queensland the round or hairy yam (*D. bulbifera*) and three introduced species of yam can occasionally be found along tropical rainforest edges and in monsoon rainforests. These are described on pages 120-121.

USES: Long yam has a very long, slender (often as thin as a pencil), crisp white tuber which tastes like potato. It can be eaten raw but is greatly improved by roasting. It was a staple food of coastal Aborigines. The Brisbane settler Constance Petrie noted: "The gins would have to dig three feet sometimes for this root ('tarm') which was very nice roasted."

D. transversa *seed pods*

Five leaf native grape (Cissus hypoglauca)

NATIVE GRAPES

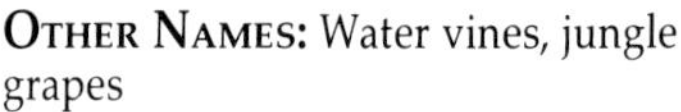

family Vitaceae

OTHER NAMES: Water vines, jungle grapes

FIELD NOTES: Native grapes are tendril-coiling vines bearing bunches of purple or black berries. They have two identifying features—berries resembling grapes, and tendrils attached to the stem on the side opposite to the leaf. Fruits ripen mainly in late summer or autumn. Leaves vary greatly between the species and provide easy identification.

Native grapes range from tall woody lianas on jungle edges to small vines of sandy or rocky soils in open forest. They are not easily confused with other vines, but not all species are edible. The following

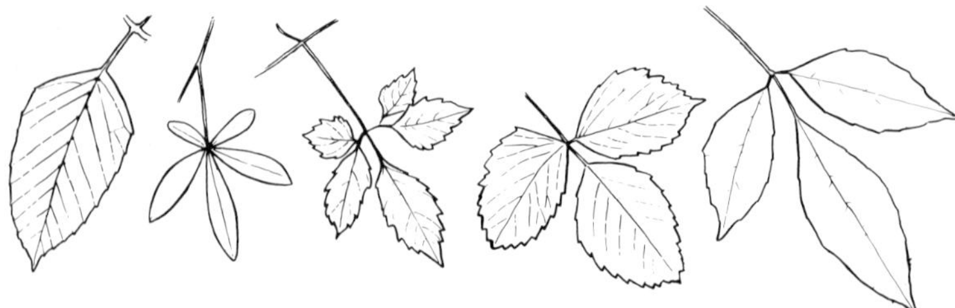

Left to right: Cissus antarctica, C. opaca, Cayratia clematidea, C. trifolia, Tetrastigma nitens

edible species can be found in eastern Australia.

Five-leaf native grape (*Cissus hypoglauca*), the only grape found south to Victoria, is one of the more common lianas of rainforest margins. The leaves are shiny and leathery, carried mainly in groups of five.

Kangaroo vine (*C. antarctica*), another very common vine of clearings and damp gullies, is found south to Narooma. Leaves are oval, with a gland on the underside at the base of each vein (see drawing), carried singly, usually with toothed margins and pubescent undersides. Young shoots are clothed in rusty hairs. This vine is often cultivated.

Pepper vine (*C. opaca*) prefers open forests and drier rainforests south to the Hunter River. It has slender leaves (less than 2.5 cm wide) in groups of three or five, which taper at their bases. This is the only grape found west of the Dividing Range.

The shining grape (*Tetrastigma nitens*) of rainforest clearings north of Gosford, is a vigorous vine with shiny leaves in threes.

The slender grape (*Cayratia clematidea*) of rainforests and open forests has soft, often pale green, deeply serrated leaves in groups of five, which are distinctively arranged. *C. eurynema* of Queensland rainforests, looks similar.

In northern Australia the most common species is the wild grape (*Ampellocissus acetosa*), a vine or squat shrub of open forests and monsoon rainforest. It has heavily veined leaves with pale undersides, arranged distinctively in groups of five to nine.

Wild grape (Ampellocissus acetosa)

Kangaroo vine (Cissus antarctica)

USES: Native grapes were widely eaten by Aborigines and colonists. They have a distinctive grape-like taste, but usually irritate the throat if eaten in any number. Colonists sometimes made them into jam, and drained drinking water from the cut stems.

Some native grapes produce insipid watery tubers, eaten by Aborigines, though these were neither important nor popular foods. Tubers of pepper vine can weigh several kilograms, making them among Australia's largest.

NATIVE GINGER

Alpinia coerulea

OTHER NAME: Common ginger, [*A. caerulea*]

FIELD NOTES: Native ginger has dark green, frond-like leaves up to 2 m tall, and resembles the various cultivated gingers grown ornamentally. It is distinguished by its bright blue fruits, about 1.5 cm wide, which consist of a brittle shell enclosing angular seeds in crisp white pulp. The fruits ripen in autumn.

There are related gingers with smaller (0.8-1 cm) blue fruits and reddish or undulating leaves. Their fruits are probably edible but information is lacking.

Native ginger grows along rainforest margins and in monsoon rainforest. It is sometimes cultivated.

USES: Aborigines ate the ginger-flavoured pulp of the fruits. It is said their paths through the forest could be detected by the trail of spat out seeds.

Young rhizomes (underground stems) taste slightly of ginger, and have been cooked as a substitute. Diced, boiled and sweetened, they make a pleasant sauce.

Two related gingers, very tall plants, 2-4 m tall, found in north Queensland lowland rainforest, have edible fruits. The native ginger (*Amomum dallachyi*) has an oval, seed-filled fruit about 3 cm long, covered by small prickles.

The native cardamom (*Hornstedtia scottiana*) has seed-filled fruits concealed within bright red bracts, resembling pineapple tops, carried at ground level.

As well, the true Asian ginger plant (*Zingiber officinale*) has been found growing wild at remote Lockerbie Scrub, Cape York. It was probably introduced by early seafarers.

S. gaudichaudiana

NATIVE ELDERBERRIES

Sambucus

FIELD NOTES: These shrubs resemble garden elderberry, having white flower clusters, pinnate leaves, and shiny 3-6 mm berries, which ripen in late summer and autumn. Yellow elderberry (*S. australasica*) has yellow fruits and grows on rainforest edges north from Gippsland. White elderberry (*S. gaudichaudiana*) has larger clusters of white fruit and prefers cooler forests and seashore scrubs between Gympie and Beachport, South Australia.

USES: Joseph Maiden wrote in 1889: "The fruit of these two native elders is fleshy and sweetish, and is used by the Aborigines for food."

S. australasica

Rainforest Raspberries

Rubus

Australia has seven native raspberries, including these five species of rainforest edges. All are prickly shrubs (called brambles) or climbers, with tooth-edged leaves and pink or white flowers with five petals. The reddish berries are 1-2.5 cm long, and ripen mainly in late summer and autumn. Australia's other raspberries are described on page 127.

Wild raspberries were widely eaten by Aborigines and colonists. Pioneer women gathered them to make preserves, pies and jams. Colonial cookery writer Mina Rawson wrote in 1894: "It is not generally known that these make a very nice preserve, insipid as they are in a raw state; they develop a good flavour when made into jam."

Queensland Raspberry

Rubus fraxinifolius

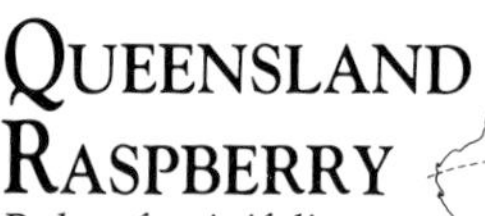

Field Notes: (Not illustrated.) This raspberry resembles the roseleaf, but the flowers and fruits are borne in large clusters.

Uses: Aborigines ate the fruits.

Roseleaf Raspberry

Rubus rosifolius

Other Names: Roseleaf bramble, forest bramble, thimbleberry

Field Notes: Leaves are soft, pinnate and bright green on both sides. White flowers are followed by distinctive hollow berries resembling strawberries. This shrub is very common along rainforest edges, in nearby paddocks and roadsides, and in gullies in eucalypt forests.

Uses: The berries are insipid, but make excellent jam.

BUSH LAWYERS

Rubus moorei, R. novae-cambriae

OTHER NAME: Bramble

FIELD NOTES: These are prickly climbers with dark green leaves, 4-12 cm long, grouped mostly in fives, on long finely-prickly stalks. Clusters of white flowers are followed by large reddish-black raspberries up to 2.5 cm long. *R. moorei*, found north of Lismore, has hairy branches; the soon to be named *R. novae-cambriae*, found south to Narooma, has smooth branches.

USES: The berries are exquisitely tangy, but contain many hard seeds.

MOLUCCA BRAMBLE

Rubus moluccanus

OTHER NAMES: Broad-leaf bramble, molucca raspberry, (*R. hillii*)

FIELD NOTES: Leaves are large and broad, 5-20 cm long, more or less heart-shaped or three-lobed, with a crinkled surface and white underside. White or reddish flowers are followed by raspberries about 1.2 cm wide. This prickly shrub or climber is common along rainforest edges as far south as eastern Victoria.

USES: The berries are insipid.

RAINFOREST KANGAROO APPLE

Solanum aviculare

FIELD NOTES: This is a large-leaved shrub with orange-red, egg-shaped fruits, 1.5-2 cm long, containing many flat seeds, 1-1.5 mm long. Flowers are purplish with yellow centres. The soft, dark green leaves are lance-shaped or deeply lobed (see leaf gallery) and measure 10-30 cm.

Rainforest kangaroo apple is a quick-growing, short-lived shrub 2-3 m tall, found in rainforest clearings in the north, and extending in the south into shaded eucalypt forests, stream margins, and (rarely) dry woodlands. It closely resembles southern kangaroo apple (*S. laciniatum*, page 133), which has larger, paler fruits.

USES: "Kangaroo apple" is a misnomer. The pulpy fruits are not eaten by kangaroos, and taste nothing like apples. The pulp is sickly-sweet, usually with a bitter aftertaste from the presence of the mildly poisonous alkaloid, solanine. In the Soviet Union and Hungary kangaroo apples (*S. aviculare* and *S. laciniatum*) are farmed for this alkaloid, which is extracted from the leaves to make contraceptive pills.

Rainforest kangaroo apple was reportedly an Aboriginal food, but the record was based on a misidentification of *S. laciniatum*. Aborigines probably ate the occasional rainforest kangaroo apple, but eaten in large amounts they are likely to cause illness.

M. australasica

RAINFOREST LIMES

Microcitrus

OTHER NAMES: See below

FIELD NOTES: The four species are all spiny shrubs or small trees with rounded or slender, sour-tasting citrus fruits, and leaves of very variable shapes, sometimes with toothed margins. Young plants can be very spiny with tiny leaves. These are uncommon plants of limited distribution in lowland rainforest.

The finger lime (*M. australasica*) has small leaves, 0.7-3 cm long, and cylindrical green, yellow, red, or purplish-black fruits, 3-10 cm long and about 1.5-2.5 cm wide. It grows from Ballina to Mount Tambourine.

The round lime or native orange (*M. australis*) has leathery leaves, 0.5-5.5 cm long, and globular greenish-yellow fruits, 2-7 cm wide, resembling rough-skinned lemons. It grows in lowland rainforest, often along streams, from Beenleigh to Gympie.

The Russell River lime (*Microcitrus inodora*) has large leathery leaves, 8-20 cm long, and egg-shaped, yellowish fruits about 5-6.5 cm long. It grows only at Bellenden Ker and Russell River, north Queensland.

The garraway lime (*Microcitrus garrawayi*) has small thick leaves up to 2.5 cm long, and greenish-yellow fruits, 5-8 cm long. It grows in vine scrubs on Cape York Peninsula.

USES: Among the more outstanding of native fruits, the native limes have exquisitely tangy, very acidic pulp, ideal for jams and drinks. Settlers gathered them for marmalade, and they are still eaten today. Colonial botanists suggested that the shrubs be brought into cultivation, and the seeds and rootstocks have been sent to the USA for experiments on disease resistance.

Snow Berry

Gaultheria hispida

Field Notes: The leaves are slender, shiny, 4–8 cm long, and the stems are furry. The fruits are 8–10 mm wide, comprising five white or pale pink lobes surrounding the seed. They ripen from March to June. Snow berry is a common shrub of Tasmanian rainforests, wet eucalypt forests, and sub-alpine shrub belts.

Uses: Aborigines ate the small fruits and settlers cooked them in tarts. Quaker Missionary James Backhouse wrote of them in 1843: "The flavour is difficult to describe, but it is not unpleasant. In tarts, the taste is something like that of young gooseberries, with a slight degree of bitterness."

Wax-berry

Gaultheria appressa

Field Notes: (Not illustrated.) Wax-berry closely resembles the snow berry. The white or pinkish lobed fruits are 7–10 mm wide and the lustrous, dark green leaves are 3–8 cm long, with paler undersides. A network pattern of veins on the leaves is a distinctive feature of both species. The fruits ripen from February to April.

Wax-berry is an 0.5–2 m tall shrub of mountainous regions, found in rainforests, sub-alpine woodlands and tall wet eucalypt forests, often near rocks, as far north as the New England Tableland.

Uses: The fruits are edible though somewhat bitter.

BOLWARRA

Eupomatia laurina

OTHER NAMES: Native guava, scented laurel, copper laurel

FIELD NOTES: This is a common shrub or small tree of the rainforest understorey resembling a lemon tree, with waxy green leaves, 6-12 cm long. The flat-topped greenish fruits are 1.5-2.5 cm wide, or up to 4.5 cm in north Queensland, with many bitter-tasting seeds. They ripen from April to June, becoming pale, soft and smelly, and toppling to the forest floor. Bolwarra is one of those very primitive plants for which the Australian rainforest is famous.

USES: Aborigines ate the sticky, sweet pulp coating the seeds.

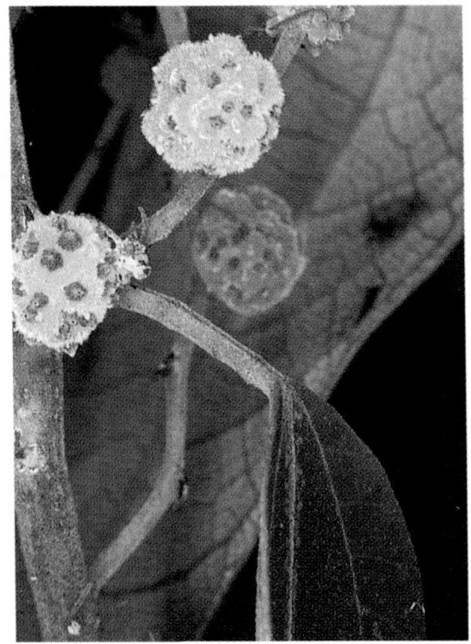

NATIVE MULBERRY

Pipturus argenteus

OTHER NAME: White nettle

FIELD NOTES: The soft leaves are 8-18 cm long, on long stalks, tapering to a point, with toothed margins, three prominent veins, and with white furry undersides. The white fruits, 5-10 mm wide, have seeds embedded in the surface; they ripen from May to July. This is a shrub or small tree of lowland rainforest edges and regrowth, damp gullies, and coral cays, from Lismore to New Guinea and the Pacific Islands.

USES: Aborigines ate the sweet but insipid fruits. In New Guinea the shoots are cooked and eaten. They are bland and fibrous.

Coast Aspen

Acronychia imperforata

Other Name: Lemon aspen

Field Notes: Leaves are leathery, oval, usually notched at the tip, distinctly segmented between the leaf base and its stalk. Fruits are orange-yellow, 1-2 cm long, with sour flesh enclosing a large stone, ripening from August to December. This is a shrub or small tree of rainforests by the sea, north of Newcastle.

Uses: The fruits taste sour and aromatic. Other rainforest *Acronychia* (for example, *A. wilcoxiana, A. suberosa*) have edible fruit. These are cream or white, four-lobed, sour and aromatic, and the leaves have jointed stalks.

Hairy Psychotria

Psychotria loniceroides

Field Notes: The leaves, stems and twigs feel distinctly furry. Clusters of small white flowers are followed by creamy-yellow fruits, 6-7 mm wide, containing one seed, which ripen throughout much of the year. This is a shrub or small tree up to 5 m tall of rainforest edges, damp eucalypt forests and seaside scrubs, north of Bega.

Uses: The watery insipid fruits are reported to be edible. The cream unripe fruits, distinguished by their hardness, taste revolting, and even the ripe fruits can irritate the throat. This is a fruit better left to the birds.

The related *P. simmondsiana* is also edible.

CRAB APPLE

Schizomeria ovata

OTHER NAMES: White cherry, white birch, squeaker, humbug

FIELD NOTES: The glossy opposite leaves are 6-15 cm long, 2-5 cm wide, usually with toothed margins. The fruits are yellowish or white, 1-1.7 cm in diameter, with sour flesh, containing one large stone, ripening from February to July. Crab apple can grow into an enormous tree, but often fruits while still shrub sized. It grows in rainforests, and sometimes in wet eucalypt forests, north of Narooma.

USES: The *completely* ripe fruits are edible but very acidic. Colonists made them into jam.

PEPPER TREE

Tasmannia insipida

OTHER NAMES: Brush pepperbush, [*Drimys insipida*]

FIELD NOTES: (Illustrated page 210.) Leaves are slender, 5-20 cm long, with wavy edges. Clusters of pinkish-white (sometimes speckled) or purplish-black berries, 1.2-2.5 cm long, furrowed down one side, ripen in autumn. This is a shrub of rainforest edges and mountain eucalyptus forests.

USES: The fruits are sweet and spongy, but the seeds taste peppery. The hot-tasting, purplish-black berries of the very similar mountain pepper (*T. lanceolata*), found from Tasmania to the Blue Mountains, were used by southern colonists as pepper.

LILLY PILLIES

Acmena, Syzygium

OTHER NAME: Lillipillies

Though not the tastiest of wild foods the lilly pillies excel in sheer visual appeal. Bunches of the shiny red, blue or purple fruits, framed by cool green foliage, are one of the rainforest's more alluring images. Australia has about 60 different lilly pillies, and nearly all have edible fruit, though some are too sour or astringent to be palatable. They are described here as a group, and the more common species are featured on the following pages.

The original lilly pilly was *Acmena smithii* but the name came to be applied to other species in genus *Eugenia* in which *smithii* was once placed. Following Hyland's 1983 revision most lilly pillies are now placed in genus *Syzygium*. "Lilly pilly" is used here to denote all *Syzygium* and *Acmena* though many are known locally as "scrub cherries", "wild apples" and the like.

Lilly pillies are small to large trees (occasionally shrubs) of rainforests (especially the edges) and wet gullies. A few northern species grow in open woodlands. They often grow beside rainforest streams and in rainforests behind beaches.

Lilly pilly leaves are always opposite, without serrated edges, and with oil glands that usually appear as pricks of light when leaves are held to the sun. Leaves of some species have a spicy clove smell when crushed (cloves are harvested from an Asian *Syzygium*).

The fruits are white, pink, red, purple, blue or black (rarely brownish or greenish). They are rounded, egg- or pear-shaped, with a single large rounded seed, and an apex with either a recessed disc (in *Acmena*) or a hollow obscured by lobes or flaps. They vary in size from about 1-9 cm diameter, and are usually carried in bunches. The flesh is usually white, spongy and watery, with a sour, fragrant, often drying taste. Some have a spicy cinnamon or clove taste.

Lilly pillies are not nutritious. The few species analysed contained mostly water (76-90 per cent) and only traces of vitamin C. They are often insipid in taste. Settlers sometimes made them into jellies and wine. Some of the northern species with large fruits (especially the lady apple) were important Aboriginal foods.

Lilly pillies are mainly tropical—only *A. smithii* extends into Victoria. Four species grow around Sydney (all described here) and 12 around Brisbane. Cape York Peninsula has more than 50 species. Apart from the species depicted on the following pages, the open forest lady apple is shown on page 148.

Lilly pillies are often grown as ornamentals, especially the species shown on the following pages.

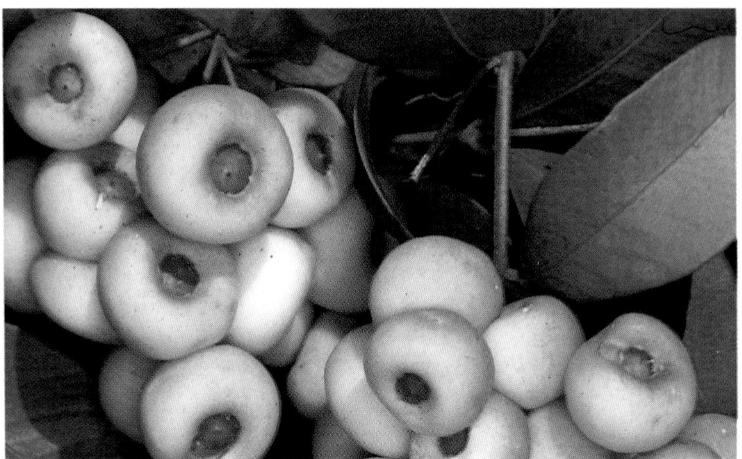

COMMON LILLY PILLY

Acmena smithii

OTHER NAME: [*Eugenia smithii*]

FIELD NOTES: The fruits are pink, white or purplish, 8-20 mm in diameter, with a characteristic depressed disc at the tip. They ripen from April to August, sometimes prolifically. The leaves are broad or slender, 2-7 cm long, sometimes much longer, and sometimes with a very tapered tip. The younger twigs are more or less square, not rounded, in cross-section.

Lilly pilly is a very common tree of coastal and highland rainforests, found south as far as Wilsons Promontory and east Gippsland. A slender-leaved form is found along streams. In the Blue Mountains it is a towering tree of rainforest creek banks, identified by its scatter of pale fruits on the forest floor in winter. By the sea it sometimes grows as a stunted, wind-sheared shrub only 1-2 m tall. It is often grown as an ornamental.

The closely related broad-leaved lilly pilly or blush satinash (*A. hemilampra*), a rainforest tree found from Port Macquarie to Torres Strait, is distinguished by its white fruits on stalks rounded in cross-section. These are also edible.

USES: Despite their alluring appearance, lilly pilly fruits are aromatic, drying and not very palatable. They were widely eaten by Aborigines. Botanist Joseph Maiden noted in 1889 that they were "eaten by the Aborigines, small boys, and birds". Colonial cookery writer Mina Rawson in 1894 claimed that the fruits make a good preserve and a good summer drink, but she may have been referring to a *Syzygium* species.

CREEK LILLY PILLY

Syzygium australe

OTHER NAMES: Scrub cherry, brush cherry, creek satinash, native myrtle

FIELD NOTES: The fruits are rounded, pear-shaped or elongated, 1.5-2.3 cm long, pink or reddish, sometimes purplish or bluish, and ripen mainly in summer and autumn. Leaves are 2.5-9 cm long with inconspicuous oil dots. This is a small tree of rainforest creek banks, seashore scrubs, dry scrubs, and creek margins in open forest. It is often grown as an ornamental.

USES: The spongy fruits have a fragrant, refreshing taste. Colonists made them into wine.

CHERRY ALDER

Syzygium luehmanii

OTHER NAMES: Riberry, small-leaved water gum, small-leaved lillipilli

FIELD NOTES: This lilly pilly is easily identified by its small (3-6.5 cm), tear-shaped leaves, and small, usually pear-shaped fruits only 8-13 mm long. These are dull (not shiny) red (rarely purple); they ripen prolifically in summer. New spring foliage is bright pink. Cherry alder grows in subtropical and littoral rainforests north of Kempsey. It is one of the most widely planted of ornamental rainforest trees.

USES: The mealy white pulp of the fruits is pleasant eating.

MAGENTA LILLY PILLY

Syzygium paniculatum

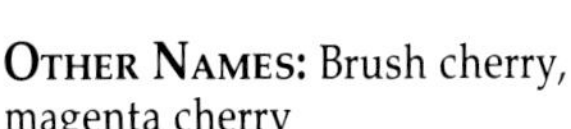

OTHER NAMES: Brush cherry, magenta cherry

FIELD NOTES: This shrub or small tree grows only in rainforests between Botany Bay to Forster, usually on sandy soil near the sea. It is often cultivated. Fruits are magenta (rarely pink, red or purple), 1.6–2.5 cm wide, and ripen mainly in autumn. Leaves are tapered, 5–9 cm long, with scattered distinct oil dots.

USES: This was the first Australian fruit to be eaten by Englishmen. Captain Cook gathered fruits at Botany Bay, and Joseph Banks wrote that "we eat with much pleasure, tho they had little to recommend them but a slight acid".

BLUE LILLY PILLY

Syzygium oleosum

OTHER NAMES: Scented satinash, blue cherry, [*S. coolminianum*]

FIELD NOTES: (Not illustrated.) The globular fruits are reddish-purple, becoming blue when ripe, 1–2 cm wide, with a distinct cavity at the tip, and ripen mainly in winter and spring. The leaves are 4–10 cm long, and densely marked with distinct oil dots, which show up as hundreds of pin-pricks when the leaf is held to the sun. The leaves smell lemony and feel sticky when crushed. This is a small tree of rainforests and coastal scrubs north from Port Kembla.

USES: The fruits are tasty, and make good jams and jellies.

WHITE APPLE

Syzygium cormiflorum

OTHER NAME: Bumpy satinash

FIELD NOTES: The big white or pinkish fruits, up to 8 cm wide, are produced up and down the trunk or on the branches of this rainforest tree. The spongy white flesh smells slightly like pears, and encloses a large round seed. The fruits ripen from October to January. Leaves are 10-20 cm long, and the creamy flowers sprout along the trunk and branches. *Syzygium branderhorstii* found on the islands of Torres Strait, also sprouts big edible fruits along the trunk. These are black, purple, red, pink or white.

USES: Aborigines ate the fruits.

DUROBBY

Syzygium moorei

OTHER NAMES: Robby, coolamon, watermelon tree, rose apple

FIELD NOTES: (Not illustrated.) Like the white apple, the durobby produces its big cream fruits along the older, leafless branches, though not on the trunk itself. The fruits are about 5-6.5 cm long, ripening in autumn. Flowers are pink or orange-red, and the leaves are glossy, dark green, large and leathery, 8-23 cm long, with paler undersides. This is a very rare tree of lowland rainforests from Casino to Canungra. It is sometimes cultivated.

USES: The fruits are edible but bland.

STINGING TREES

Dendrocnide

D. moroides

OTHER NAMES: Stingers, Gympies, fibrewoods

FIELD NOTES: These are shrubs or trees mainly of rainforest edges, with stinging hairs and fruits resembling raspberries or mulberries. At least three of the five species are edible.

Gympie stinger or stinging bush (*D. moroides*) is a shrub or small tree, found from Lismore (where rare) to north Queensland. It has very big, 10–22 cm long, broad, flat, pale leaves, with deeply serrated edges which are densely clothed in stinging hairs on both sides. The leaf stalk is attached, not to the leaf edge, but to the back (as on a garden nasturtium) near the base. The insipid tasting fruits are dark pink or purplish-red.

Giant stinging tree (*D. excelsa*), found from Bega to Gympie, is common as a small tree of rainforest clearings, but grows into a towering giant. Leaves are like those of the Gympie stinger, but are finely toothed, hairier below than above, and have the stalk attached to the rear of the leaf, which is more or less heart-shaped. The fruits are white or pink.

D. photinophylla

The very different shiny-leaved or mulberry-leaved stinging tree (*D. photinophylla*), found from Newcastle to north Queensland, is a small tree with shiny, bright green leaves, 5–18 cm long, which are only slightly if at all toothed. The tangy fruits are creamy white. A few scattered stinging hairs adorn the leaf undersides and fruits.

USES: The fruits are edible, but should first be rubbed in a cloth to remove any stinging hairs.

NATIVE FIGS

Ficus

Although the Biblical fig tolerates a dry climate, most of Australia's 42 native figs are rainforest or river bank trees. They all produce figs—rounded, soft-skinned fruits full of tiny seeds—of varying quality, ranging from sweet, succulent and delicious to dry, gritty and tasteless. None are poisonous.

Sandpaper figs (illustrated pages 83, 147) are easily recognised by their sandpapery leaves which Aborigines used to polish their weapons. Sandpaper figs are shrubs or small trees with soft figs that can be eaten whole.

The larger groups of figs are those that can be eaten whole. These are often majestic jungle trees with enormous probing surface roots that coil about rocks, stumps or other trees. The leaves are often leathery, on long stalks, and the young shoots are enveloped in a spike-like stipule. Milky sap bleeds from broken stems. The figs consist of a soft sweet flesh surrounding a dry seed mass not usually worth eating. The flesh, though usually insipid, tastes fig-like. Some kinds of figs bear the fruits along the trunk or branches.

Aborigines probably ate most kinds of figs, and the fibrous bark of some were used to make nets.

Only a handful of fig species are illustrated here. Of the remainder, some are yellow, black or white when ripe. Native figs are mostly eastern and northern trees, with 11 species occurring south to Brisbane, five to Sydney and one to Gippsland. They are common in tropical rainforests, monsoon rainforests, vine scrubs, and jungles lining creeks and rivers. The desert fig (illustrated page 175) is the only outback species.

WHITE FIG

Ficus virens

OTHER NAME: Banyan

FIELD NOTES: (Not illustrated.) This sprawling tree has aerial roots which sprout from the undersides of the branches and grow downwards, eventually reaching the ground and forming a series of smaller trunks. Leaves are deciduous, 6-19 cm long, on stalks 2.5-7 cm long. The figs are white, pink or purplish, often spotted, 0.8-1 cm wide. White fig grows in lowland rainforests, monsoon rainforests and vine scrubs, often close to the sea.

USES: Aborigines ate the small figs. The Chinese in Darwin cook the new season's leaves as a vegetable.

F. macrophylla

MORETON BAY FIG

Ficus macrophylla

OTHER NAME: Black fig

FIELD NOTES: This giant rainforest tree, found north of Nowra, is often planted in parks or left in paddocks when rainforests are cleared. It usually begins life by strangling other trees. The leathery leaves are 8–25 cm long, 5–8 cm wide, usually with rusty undersides. The purplish-red figs are 2–2.5 cm wide with white or yellow spots. They ripen year-round but mainly in autumn. The gigantic buttressed roots trail over the ground.

USES: Aborigines ate the sweet flesh of the figs, and made string from the bark. Most of the fruit consists of gritty seeds.

STRANGLER FIG

Ficus watkinsiana

FIELD NOTES: The most common and spectacular of strangling trees, this is a huge, often emergent tree of rainforests north of Newcastle. The fallen figs provide easy identification. They are very large, 3–6 cm long, dark purplish with spots, nearly always with a nipple at the end, ripening year-round.

USES: The succulent figs are very tasty.

Fruits of the strangler fig

CLUSTER FIG

Ficus racemosa

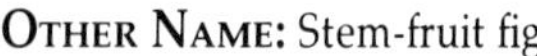

OTHER NAME: Stem-fruit fig

FIELD NOTES: The red figs are 4-7 cm wide, produced in clusters along the trunk of a tall deciduous tree with oval leaves, 6-20 cm long, and a pale grey, sometimes buttressed, trunk. The figs ripen at any time of year.

Cluster fig grows mainly along lowland streams and rivers, but also in seashore scrubs and monsoon rainforest, in Australia, New Guinea and southern Asia. It is sometimes grown in parks. Several other kinds of fig trees produce figs along the trunk, and these too are considered edible.

USES: Explorer Ludwig Leichhardt encountered cluster fig trees during his explorations of the Burdekin River in central Queensland in 1845. His journal notes:

> These trees were numerous, and their situation was readily detected by the paths of the natives leading to them: a proof that the fruit forms one of their favourite articles of food.

Leichhardt often sampled the figs, and noted they had an "agreeable flavour", but suffered severe indigestion after sometimes eating too many.

The figs are very pleasant eating. In the Port Curtis district, according to C. Hedley writing in 1888, the tree "furnishes the settlers with excellent jelly" and yields "to the blacks edible fruits". In India the young shoots and green figs were cooked and eaten.

Some Aboriginal groups in the Northern Territory make a preparation from the inner wood to treat diarrhoea.

CREEK SANDPAPER FIG
Ficus coronata

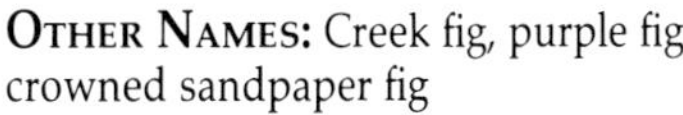

OTHER NAMES: Creek fig, purple fig, crowned sandpaper fig

FIELD NOTES: Leaves are dark green, alternate, often asymmetrical at the base, sandpapery above and hairy below. Ripe figs are soft, purplish-black, 1.5-3 cm long, very hairy, and mainly sprout on the twigs and branches from January to June. This is a small tree of rainforest creek banks and gullies south to Mallacoota Inlet, sometimes found extending along creeks into open forest.

USES: Some crops of the figs are sweet and flavoursome, others dry and insipid. The leaves make excellent sandpaper.

WHITE SANDPAPER FIG

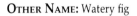

Ficus fraseri

OTHER NAME: Watery fig

FIELD NOTES: (Not illustrated.) Leaves are dark green, 6-13 cm long, mainly alternate, symmetrical at the base, with a rough upper surface but not hairy below. The figs are rounded and hairless, about 1-1.5 cm wide, turning yellow, red, then black when ripe. This is a small tree of rainforests along creeks and rivers and behind beaches, and of drier vine scrubs, north of Newcastle.

Other species of sandpaper fig, distinguished by their mainly opposite leaves, are described on page 147.

USES: The figs are edible but somewhat insipid.

Palms

family Palmaceae

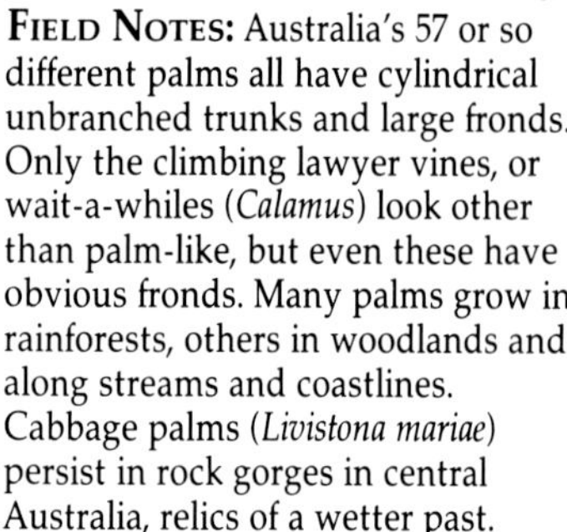

Field Notes: Australia's 57 or so different palms all have cylindrical unbranched trunks and large fronds. Only the climbing lawyer vines, or wait-a-whiles (*Calamus*) look other than palm-like, but even these have obvious fronds. Many palms grow in rainforests, others in woodlands and along streams and coastlines. Cabbage palms (*Livistona mariae*) persist in rock gorges in central Australia, relics of a wetter past.

Uses: The growing bud ("cabbage" or "heart") atop the trunk of any palm may be eaten raw or cooked, though it kills the tree. This food was known to eighteenth century mariners, and Captain Cook sampled palm cabbages at Endeavour River, including one that tasted "exquisitely sweet". Australia's first colonists harvested palm cabbages to supplement meagre rations, and the cabbage palm (*L. australis*) takes its name from this time. One early writer described it as having a "white and rather spongy texture, which possesses the sweet taste of the Spanish chestnut and is often eaten by whites as well as by the Aborigines". The hearts were important foods of Aborigines.

In South America palm cabbages are very popular, and six million are harvested each year, many for the canned export market. Some palm species are now overharvested to the point of extinction.

The walking stick palm (*Linospadix monostachya* illustrated) of rainforests in northern New South Wales and southern Queensland, sprouts strings of edible red fruit in autumn. Lawyer vines also produce scaly edible fruits.

Cunjevoi

Alocasia macrorrhizos

Field Notes: Leaves are heart or arrow-shaped, 0.5-2 m long, arising from a thick trailing rhizome (root-like stem). Strongly scented yellowish flowers are followed by inedible red berries.

Cunjevoi grows in rainforests and moist gullies. It is sometimes cultivated as "spoon lily".

Uses: The cunjevoi is one of the most dangerously poisonous of Australian plants, and children have died from nibbling the leaves and stems. The rhizomes, though poisonous, are very starchy, and Aborigines soaked and baked them to remove the toxin. Settler Tom Petrie noted that the rhizomes were soaked lengthily, pounded, made into cakes and roasted. According to another report young "bulbs" were scraped, baked, pounded and baked again, the process being repeated eight or ten times. Aborigines no longer harvest this dangerous plant.

Linospadix monostachya *centre right, and* Alocasia macrorrhizos *below*

D. antarctica

TREE-FERNS

Cyathea, Dicksonia

FIELD NOTES: Tree-ferns have a crown of large fronds arising from the top of an unbranched trunk. They grow in rainforests and wet eucalypt forests, usually along stream banks or on wet mountain slopes. Australia has 11 species of *Cyathea* and two *Dicksonia*. Rough tree-fern (*C. australis*) and soft tree-fern (*D. antarctica*) are the most common species. The former has croziers (curled shoots) covered at their bases in shiny, flattened, ribbon-like scales; the soft tree-fern has croziers covered in coarse, brittle hairs.

USES: The upper trunk of the tree-fern contains a core of white starch which was an important Aboriginal food, especially in Tasmania. Missionary James Backhouse in 1843 described the destruction wrought in pursuit of this food:

> In passing through a woody hollow, we saw many of the tree-ferns, with the upper portion of the trunk split, and one half turned back. This had evidently been done by the Aborigines, to obtain the heart for food, but how this process was effected, I could not discover; it must certainly have required considerable skill.

Other early witnesses describe the core as a turnip-like substance as thick as a man's arm, tasting variously bitter, sweet, astringent, or "like a bad turnip". Some species tasted better than others, and in Tasmania it was said the Aborigines ate one kind of tree fern only with kangaroo and other meats, "whilst that from the [*Cyathea*] was considered so good that it might be partaken of alone". The starch was usually roasted but could be eaten raw.

Tree-fern pith was probably a mainstay of some tribes. A sample of soft tree-fern yielded 61 kilocalories per 100 grams, about twice that of turnips or pumpkins. Removing the core kills the tree and should not be attempted today.

The uncurled fronds of tree-ferns, called croziers, are also edible, and have a juicy, slimy, sometimes bitter taste.

NATIVE TAMARIND

Diploglottis cunninghamii

OTHER NAME: [*D. australis*]

FIELD NOTES: Native tamarind is one of Australia's more distinctive rainforest trees. The huge furry leaves (among the largest of any rainforest tree) are in large fronds atop a slender, often unbranched trunk. The fruits are yellowish or orange-brown furry capsules with one to three lobes, each about 1-1.5 cm wide. The lobes split when ripe to expose the apricot-flavoured pulp surrounding a large seed. The fruits can be found on the forest floor in summer, along with the furry fallen leaves.

The native tamarind is a small or large tree of subtropical rainforest, found mainly along edges and clearings.

USES: The fruits have a juicy sour flavour, ideal for drinks and jam. Colonists made them into jelly "with a pleasant tang of the wild about it".

There are other native tamarinds with large pinnate leaves and tangy edible fruits in pods that split open when ripe. The wild tamarind (*D. diphyllostegia*) is a north Queensland species of very similar appearance. The small-leaved tamarind (*D. campbellii*), a very rare tree of the Queensland-New South Wales border rainforests, has big, 3-6 cm wide, rounded fruits with bright red pulp. The Atherton tamarind (*D. bracteata*) of the Atherton Tableland has three-lobed fruits 2.5-4.5 cm wide with orange pulp. The corduroy tamarind (*Arytera lautereriana*) of Queensland rainforests has slender pinnate leaves and reddish three-lobed fruits with pale yellow pulp. The true tamarind, an Asian tree with long bean-like pods, is featured in Chapter 7.

BROWN PINE

Podocarpus elatus

OTHER NAMES: Plum pine, she pine, yellow pine, Illawarra plum

FIELD NOTES: Brown pine is unusual among rainforest trees in having very tough, narrow, sharp-tipped leaves, 4-8 cm long, about 1 cm wide. The bark is brown and finely furrowed. The fruit has two segments—a hard inedible seed about 1 cm wide, and a larger basal portion with juicy purple flesh. Fruits ripen from March to July and soon fall to the forest floor.

Brown pine is a tall tree of dense subtropical, riverine and seashore rainforests. It is often planted in parks where it fruits prolifically. The timber is ideal for furniture and boat planking, but most wild stands were felled or cleared last century.

USES: The dark juicy pulp of the fruit tastes rich and sweet, but has a very resinous-tasting central core, which is best avoided. In southern New South Wales it was esteemed by Aborigines and settlers as one of the best of wild fruits, but in Queensland, where the choice of bush fruits is much wider, it was not highly regarded. Colonial botanist Joseph Maiden wrote in 1898 that the fruits "are known to our small boys as plums or damsons". A sauce made from them today is served in wild food restaurants.

BURDEKIN PLUM

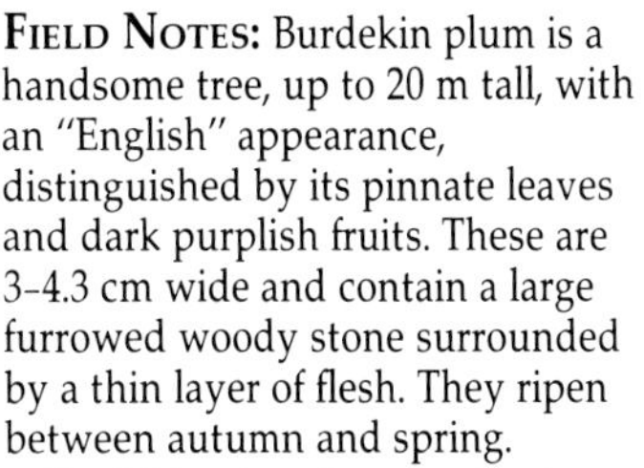

Pleiogynium timorense

FIELD NOTES: Burdekin plum is a handsome tree, up to 20 m tall, with an "English" appearance, distinguished by its pinnate leaves and dark purplish fruits. These are 3-4.3 cm wide and contain a large furrowed woody stone surrounded by a thin layer of flesh. They ripen between autumn and spring.

Burdekin plum favours riverine rainforest, jungle gullies, dry vine scrubs, coastal scrubs on sand, rocky hillsides, volcanic craters and old lava flows. It grows from Gympie north to the Philippines, west to Indonesia (hence *timorense*), and east to the Cook Islands. It is sometimes planted in parks.

USES: The Burdekin plum produces large crops of very sour and astringent fruits which only become palatable to western tastes some days after falling to the ground. Joseph Banks noticed this after collecting fruits at Endeavour River: "these when gathered off from the tree were very hard and disagreeable but after being kept for a few days became soft and tasted much like indifferent Damsons."

The fruits were widely eaten by Aborigines, as noted by several early observers. The pulp, with the addition of pectin or lemon juice, makes an excellent tangy jam, and trees producing high quality fruits are prized in north Queensland by Aborigines and whites alike.

Australia's armed forces have investigated the fruit as a potential survival food but it scored poorly as a source of kilojoules, protein, vitamins and minerals.

Davidson's Plum

Davidsonia pruriens

Other Names: Ooray, sour plum

Field Notes: The large fronds of saw-edged leaves are clothed in irritating hairs. The purplish fruits are 3-5 cm long, contain two flat seeds, and hang from the trunk and branches. This is a small, slender tree of rainforests north of Mullumbimby. It is sometimes cultivated.

Uses: The very sour fruits are exquisitely tangy, and are rated among the best of native fruits, though they really need to be stewed or made into jam to be properly appreciated. Sauces made from the fruits (gathered from cultivated trees) are served in wild food restaurants.

Black Plum

Planchonella australis

Other Names: Wild plum, black apple

Field Notes: The plum-like fruits are 4-6 cm long, ripen in late spring and summer, and contain three to five long shiny brown seeds, which separate cleanly from the soft pulp. They are found on the ground beneath a tall rainforest tree with slender leaves, 6-14 cm long, and a fluted trunk, which oozes milky sap when cut.

Uses: Aborigines ate the juicy fallen fruit and settlers gathered them for preserves. The pulp is bland but pleasant, though usually full of maggots. The seeds are also said to be edible. Other *Planchonella* are edible.

BLUE QUANDONG

Elaeocarpus grandis

OTHER NAMES: Silver quandong, brush quandong, blue fig, coolan

FIELD NOTES: Quandong is a tall majestic rainforest tree with open, almost horizontal branches that carry slender leaves with finely serrated edges. The blue fruits, about 2-3 cm in diameter, consist of a thin layer of sour green flesh surrounding a large, spherical, very nobbly stone. They ripen from May to January.

Blue quandong grows mainly beside streams in dense coastal and mountain rainforests. It is unrelated to desert quandong (illustrated page 177) but has similar stones.

USES: Quandong fruits in season are favourite fare of fruit pigeons, especially wompoos, top-knots and purple-crowned pigeons. Laden trees can be located by the beat of flapping wings. Aborigines once gathered the fallen fruits, sometimes kneading them with water into an edible paste. They are sour, insipid, and not very nutritious. They contain no thiamine and only traces of vitamin C.

There are at least two related trees with serrated leaves and blue edible fruits. The blueberry ash (*E. reticulatus*), a shrub or small tree of damp forests from Tasmania to south Queensland (common around Sydney), has egg-shaped or rounded fruits about 1 cm long, with a meagre layer of barely edible pulp. This plant is often cultivated.

The Arnhem Land quandong (*E. arnhemicus*) of scrubs in far northern Australia, has egg-shaped fruits 1.2-1.5 cm long.

On the Atherton Tableland the Atherton oak (*Athertonia diversifolia*) produces a very large oval blue fruit containing a pock-marked, lens-shaped seed, the kernel of which is delicious.

Macadamia Nuts

Macadamia

Other Names: Queensland nut, bauple nut, bush nut

Field Notes: There are two species, both small rainforest trees with hard slender leaves and bony nuts. The smooth macadamia of southern Queensland (*M. integrifolia*, illustrated) has blunt-tipped leaves with smooth or spiny margins, leaves mostly in threes, and smooth round nuts. The uncommon rough-shelled macadamia (*M. tetraphylla*), found from Lismore to Mount Tambourine, has spinier pointed leaves, mostly in fours, and rougher nuts.

The similar but inedible maroochy nut (*M. ternifolia*) is distinguished by its sharp-tipped leaves in threes, and bitter nuts.

Uses: The exquisite macadamia nut, Australia's proud contribution to world agriculture, was actually first cultivated by Americans. Australian botanists were urging its development in 1900, but seeds were already on their way to Hawaii, where American farmers developed a multimillion dollar industry long before Australia's CSR was spurred into production in 1963. Many varieties now grown on southern Queensland farms bear Hawaiian names.

Though Australians were slow to develop macadamias commercially, the smooth macadamia was widely planted in parks and gardens. Many trees from the nineteenth century still bear nuts today. As a child I remember cracking the nuts with hammers or bricks, using ever-widening cracks in concrete to secure the nuts. Aborigines relished the oily nuts, and nineteenth century foresters were forbidden to fell trees.

CANDLENUT

Aleurites moluccana

OTHER NAME: Carie nut

FIELD NOTES: Candlenut is a distinctive large-leaved tree of rainforest margins, regrowth, and jungle gullies. The leaves are shiny green above, paler below, 12-30 cm long, on stalks of similar length, often with three or five lobes and resemble maple leaves. Young leaves are clothed in a white bloom which gives the trees a silvery cast in spring.

Clusters of white fragrant flowers are followed by green capsules, 4-8 cm wide, enclosing one to three oily white seeds, each enclosed in a hard ridged shell, 2.5-3.5 cm wide.

The trees are widely planted in parks in Queensland and New South Wales and the cultivated trees bear heavily.

USES: Candlenut was a popular food of Polynesian seafarers, who carried it to many Pacific islands, including Hawaii, where it is proclaimed the state tree. On some islands the seeds were skewered on sticks and lit as candles, hence the name. The kernels are 50-60 per cent pure oil, and burn freely.

North Queensland Aborigines roasted candlenuts in a slow fire, and when the shells cracked the kernels were fit to eat. Raw nuts can be emetic.

The kernels are an excellent source of thiamine, in one test yielding over 4000 micrograms per 100 grams—an Australian wild food record. They have a pleasant taste but are much blander than macadamias and other nuts. Aborigines used the nut oil to fix pigments to spears.

MORETON BAY CHESTNUT

Castanospermum australe

OTHER NAME: Black bean

FIELD NOTES: The pinnate leaves smell of cucumber when crushed. Big orange or red pea flowers are followed in autumn by dangling pods, 12-25 cm long, containing three to five big seeds, 3-6 cm long. This is a handsome tree found mainly on river and creek banks in rainforests.

USES: The starchy seeds were an Aboriginal staple, eaten after careful preparation to leach out a poisonous alkaloid. The seeds were sliced finely, soaked in streams, then cooked as meal. Colonists sometimes prepared and ate the starch.

H. pinnatifolia

RED BOPPLE NUT

Hicksbeachia

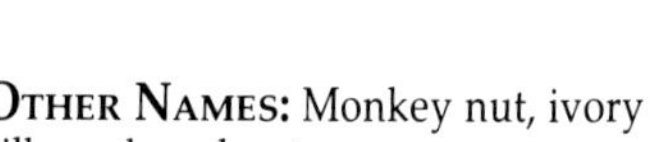

OTHER NAMES: Monkey nut, ivory silky oak, red nut

FIELD NOTES: These small, slender, rainforest trees have big, leathery, serrated-edged fronds and shiny red fruits, 3-5 cm long, which hang in clusters in spring and summer. Each fruit has a leathery rind enclosing an edible seed. *H. pilosa* grows in north Queensland and the rare *H. pinnatifolia* in the south, from Mount Tambourine to Nambucca Valley.

USES: Bopple nuts are related to macadamias, but contain less oil (about 13 per cent) and are not as flavoursome.

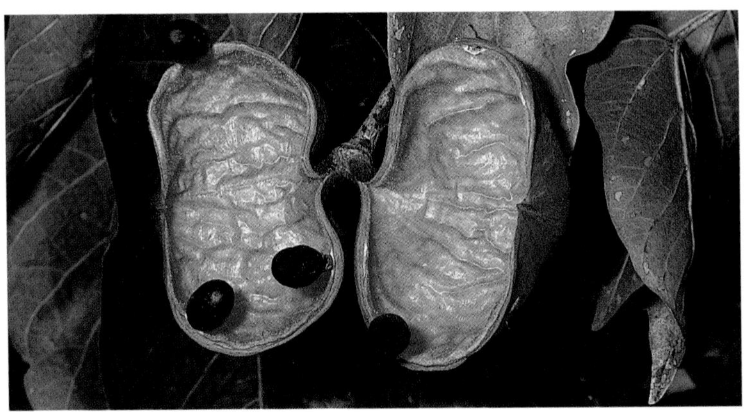

PEANUT TREE

Sterculia quadrifida

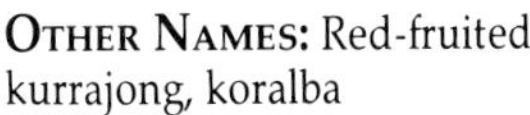

OTHER NAMES: Red-fruited kurrajong, koralba

FIELD NOTES: This deciduous tree has large, 8–20 cm, dangling, oval, heart-shaped or lobed leaves on long stalks, greenish bell flowers, and very distinctive red leathery pods, 4–7 cm long, which split to expose shiny black seeds. They ripen from April to December. Peanut tree grows in monsoon, riverine and dry rainforests, and in seashore scrubs, south to Coraki. It is common around Darwin.

USES: The peeled, peanut-flavoured seeds are a delicious Aboriginal snack food. The leaves were sometimes cooked with meats as sweetening herbs.

BUNYA NUT

Araucaria bidwillii

FIELD NOTES: (Illustrated page 14.) Bunya pines are towering trees with straight trunks, a rounded crown, and very slender branches with upturned ends. Leaves are dark green, leathery, sharp-tipped, 1–6.5 cm long. The green, rounded cones, about 30 cm long, contain many starch-filled seeds about 4–5 cm long. Bunyas grow as rainforest emergents from the Conondale Ranges north to Gympie and westwards, and in north Queensland. They are often planted in parks.

USES: Aborigines came from afar to feast upon the starchy seeds at huge social gatherings. The seeds were eaten raw or baked; they are delicious boiled. Supermarkets sometimes sell them.

AUSTRALIAN CASHEW NUT

Semecarpus australiensis

OTHER NAMES: Tar tree, marking nut

FIELD NOTES: The Australian cashew is a tree with large, dark green leaves and smooth brown bark, easily identified by its unusual fruits. From afar it looks like a mango tree. The leaves are alternate, 10-30 cm long, 4-12 cm wide, prominently veined, with paler undersides. The stems ooze milky sap, which turns black upon exposure to air. Small cream flowers are followed in summer by fruits, 2-4 cm wide, which consist of a seed inside a leathery pod attached to an orange or red fleshy base.

The cashew tree grows in open forest or scrub behind beaches and in rainforest bordering lowland rivers and streams. It ranges into the Torres Strait Islands to New Guinea.

USES: The sap of the bush cashew tree is extremely irritating, and Aborigines exercised great care when gathering and preparing the seeds. Some tribes coated their hands in clay beforehand. Before eating, the seeds were roasted over a fire, creating a smoke that was considered poisonous. The seeds taste like cultivated cashews, to which they are related. The completely ripe fruits are also edible, if gathered from the ground and peeled (and preferably baked). Explorer Ludwig Leichhardt sampled the fruit at Raffles Bay, finding it "extremely refreshing; the envelope, however, contained such an acrid juice that it ate into and discoloured my skin, and raised blisters wherever it touched it". Severe skin poisonings still occur today.

P. debilis

Native Plantain

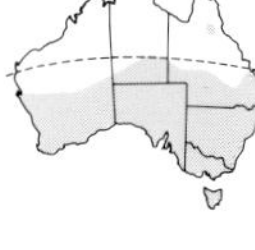

Plantago

Other Name: Sago-weeds

Field Notes: Plantains are herbs consisting of a basal cluster of slender strap-like or broad spoon-shaped leaves, which have three prominent longitudinal veins. The leaves are often furry and their margins are sometimes sparsely toothed. Minute greenish flowers and seeds are produced along slender upright stalks.

Australia's 24 species grow in grasslands, woodlands, forests and alpine meadows.

Uses: When heavy rains drench the seeding stalks of plantains, the tiny seeds swell into balls of jelly, which stick to passing animals and thereby help spread the seeds. The jelly is mucilage, a traditional herbal cure for constipation. European herbalists administer seeds of African and Indian plantains, which swell in the stomach and add bulk to the faeces.

Colonists in western New South Wales made "sago" puddings by adding boiling water and sugar to wild plantain seeds. The species used was probably *P. cunninghamii* known to this day as "sago-weed".

The use of plantains may have been learned from Aborigines—there is a report from north-western New South Wales that Aborigines there bruised the seeds of "Native sago", identified as *Plantago varia* (nowadays called *P. debilis*), to make "a kind of porridge".

The French botanist Labillardiere in 1793 ate the leaves of a Tasmanian plantain, and reckoned it "among the most useful plants, which this country affords for the food of man . . . the salad furnished by the leaves of this plant, which were very tender, was highly relished by all the company".

Scrub Nettle

Urtica incisa

Field Notes: Leaves are triangular and opposite, 5-12 cm long, with serrated margins and stinging hairs. Stems are very hairy and the flower spikes are greenish and lumpy. The introduced small nettle (*U. urens* also edible), is similar but has smaller, more rounded leaves. Scrub nettles grow in moist forests along trails and streams, rainforest margins, and in cow paddocks.

Uses: Aborigines along the Murray River in South Australia ate nettles, first baking the leaves between heated stones. They make a very tasty vegetable. Colonists boiled them to make a tonic for "clearing the blood".

Common Sowthistle

Sonchus oleraceus

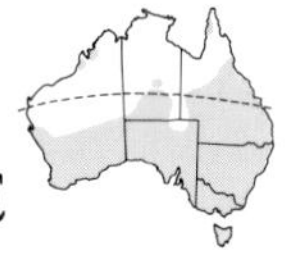

Field Notes: This familar herb has yellow, dandelion-like flowers, soft-lobed leaves with tiny spines along the margins, and hollow stems up to 1.5 m tall that ooze milky sap when cut. Sowthistle is a common garden weed, found also in fields, beaches and river banks throughout much of the world. It is wrongly believed to be an introduced plant in Australia.

Uses: Aborigines in Victoria and South Australia were very fond of the tender but bitter-tasting leaves, which taste like endive, a closely related plant. Sowthistle was also eaten by Asian and African peasants, and by the explorers Stuart and Grey.

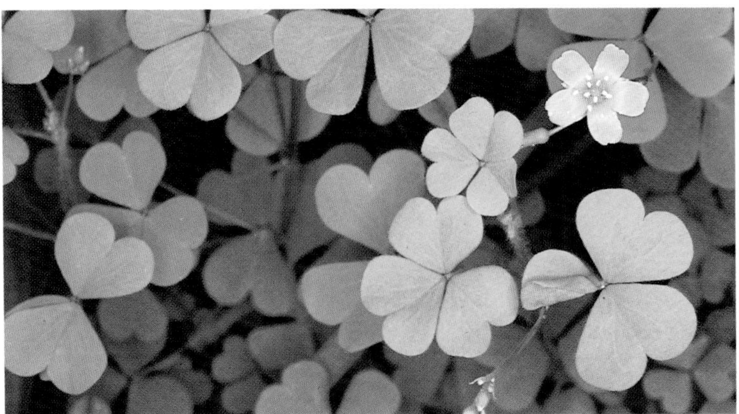

O. corniculata

YELLOW WOOD SORREL

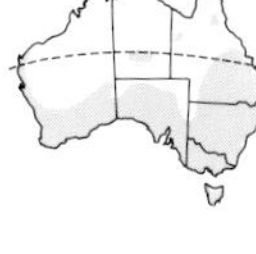

Oxalis

OTHER NAMES: Sour-weed, sour grass, clover sorrel

FIELD NOTES: Yellow wood sorrel is a soft herb easily identified by its sour-tasting, clover-like leaves (each consisting of three heart-shaped leaflets), and by its yellow flowers with five petals, 2-12 mm long.

There are several Australian species. *O. corniculata* is an introduced weed of lawns and gardens, which many botanists do not distinguish from the native *O. oxalis, O. perrenans* and *O. radicosa* found in deserts, woodlands, coastal dunes, and along rainforest edges. As well, there are many edible exotic *Oxalis* found growing as weeds, with larger flowers of varying colours.

USES: The very sour, lemony leaves of wood sorrel were sometimes eaten by Aborigines and early settlers. Quaker missionary James Backhouse wrote in 1843: "This little plant, which displays its lively, yellow blossoms on almost every grassy spot in the colony [of Van Diemen's Land] . . . is very pleasant eaten raw, to allay thirst; and made into tarts, it is almost equal to the barberry." Victorian goldminers ate the plant to prevent scurvy, and in New South Wales it was considered to be "an excellent salad herb".

Some of the native wood sorrels have small taproots which Aborigines in South Australia dug as food. Among Adelaide tribes, according to one early writer, "the root most sought after is a highly nutritious *oxalis* resembling the small carrot and tasting like coconut". Another observer said the roots were dug with pointed sticks about five feet long.

SMALL TUBER-BEARING LILIES

Small tuberous lilies with pretty flowers and grass-like leaves can be found in grassy woodlands, open forests and heaths. These lilies often grow together in large numbers, and Aborigines dug their tiny tubers for food.

The tubers are an adaptation to Australia's harsh seasonal climate. Most of these lilies sprout leaves and flowers only during favourable seasons, later withering and surviving only as underground tubers (see Chapter 4).

In southern Australia, ground lilies grow alongside orchids and murnong, sometimes in such dense colonies that there appears to be an underground layer of tubers a few centimetres beneath the surface. Aboriginal women turned the soil with sticks and reaped a bountiful harvest. Despite their small size, these tubers could have served as staple foods, especially the starchy roots of bulbine lily and milkmaids.

The tuberous lilies all have small flowers, less than 2.5 cm wide, with three or six petals (properly called tepals). These are purplish, pink, blue, yellow or white. In temperate Australia most lilies flower from late winter to early summer; the tropical lilies flower during the wet season. Some lilies (fringed lilies, vanilla lilies) have short-lived flowers that open only for half a day; the flowers of milkmaids and early nancy stay open for several days. When the flowers have closed, these lilies are very difficult to find in their grassy habitat. The leaves are always slender and grass-like.

The tiny tubers vary in shape and form (see tuber gallery). Some taste crisp and sweet, others bitter and slimy whilst the older, darker tubers in any crop taste bitter. Most of the tubers are edible raw, and cooking does little to improve their flavour. Some lilies produce only one tuber, others from five to 20 or more tubers.

Most of these lilies belong in family Liliaceae, apart from the pink swamp lily and arda (both Commelinaceae). The proper identification of some species requires clarification. The pink swamp lily, as presently defined, actually consists of two species, one in northern Australia with wet buds and stamens arranged bilaterally, and a smaller form in eastern Australia (illustrated page 105) without these features. The robust, central Queensland form of the bulbine lily, found on heavy black clay soils, is probably a unique species. Some specimens of chocolate and vanilla lilies appear to be intermediate between the two species.

All of the lilies that were known Aboriginal foods are described here, but there are no doubt many other edible species, for example, blue stars (*Chaemascilla corymbosa*—see tuber gallery) in southern Australia and the blue-flowered *Caesia setifera* in the tropics. When identifying lilies, refer to the tuber gallery and to page 20—the tuber shapes are useful identifying features.

W. dioica

MILKMAIDS

Burchardia umbellata

OTHER NAME: Star-of-Bethlehem

FIELD NOTES: The one to ten flowers sprout in a cluster atop a wiry stalk, 15-65 cm tall. The white (rarely pale pink) flowers have six concave petals and a reddish or purplish three-sided pistil in the centre. The one to four leaves are up to 60 cm long. Three to ten slender, carrot-shaped, whitish tubers are clustered at the base of the plant.

Milkmaids is common in woodlands and heaths, often growing on hillsides or in sandy soggy soils. Flowers open from late winter to summer.

USES: Aborigines ate the crisp juicy tubers which have a pleasant raw potato taste.

EARLY NANCY

Wurmbea

OTHER NAMES: Harbinger-of-spring, blackman's potatoes, star lily, [*Anguillaria dioica*]

FIELD NOTES: Flowers are white, sometimes tinged pink, green or yellow, and often have a band or pair of spots on each petal. The flowers sprout along a single, often zig-zag stalk, appearing mainly in late winter and spring. The bright green leaves are borne along the stem; the leaf bases sheath the stem. These are small lilies of woodlands and open forest.

USES: Aborigines ate the tiny rounded tubers of some species, but most seem unpalatable, including the common *W. dioica*.

A. milleflorum

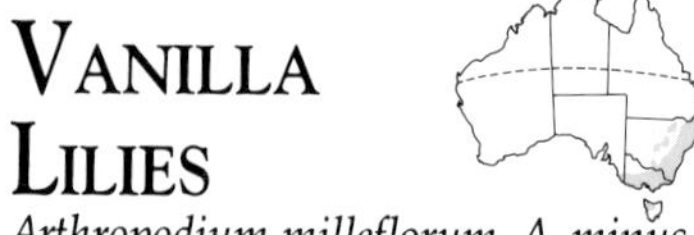

Vanilla Lilies

Arthropodium milleflorum, A. minus

Field Notes: These lilies are identified by the jointed stems attaching the flowers to the main stalk, which is up to 60 cm tall. The inconspicuous mauve, lilac or white, drooping flowers are 6-13 mm wide, with hairy stamens, and open mainly in spring and summer. Tubers are egg-shaped, 1.5-3 cm long, sprouting at the base of the small vanilla lily (*A. minus*), and along the lengths of the roots of the pale vanilla lily (*A. milleflorum*). These lilies grow in forests, woodlands and grasslands—they are inconspicuous and easily overlooked.

Uses: Aborigines ate the juicy, sweetish or bitter tubers.

C. parviflora

Pale Grass Lily

Caesia calliantha, C. parviflora

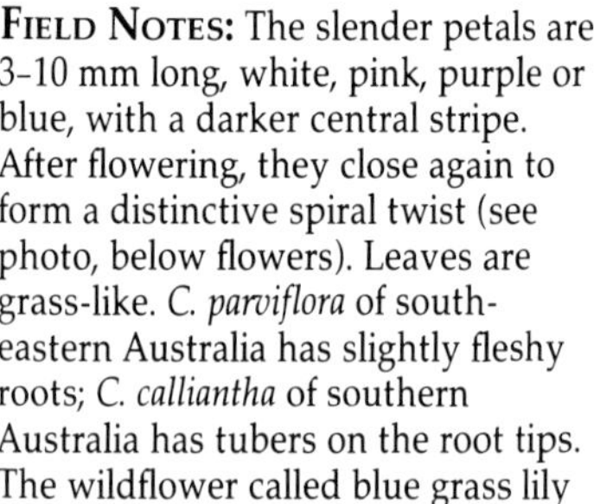

Field Notes: The slender petals are 3-10 mm long, white, pink, purple or blue, with a darker central stripe. After flowering, they close again to form a distinctive spiral twist (see photo, below flowers). Leaves are grass-like. *C. parviflora* of south-eastern Australia has slightly fleshy roots; *C. calliantha* of southern Australia has tubers on the root tips. The wildflower called blue grass lily (*C. vittata*) is now considered a variety of *C. parviflora*.

Grass lilies grow in heaths and woodlands, where they are easily overlooked.

Uses: Aborigines apparently ate the roots of both species.

H. hygrometrica

BULBINE LILY

Bulbine bulbosa

OTHER NAMES: Golden lily, leek lily, native leek, wild onion, yellow onion weed, [*Bulbinopsis bulbosa*]

FIELD NOTES: The bright yellow flowers are clustered along stalks 20–85 cm tall. The bright green leaves are hollow and succulent, like onion leaves. The single, dark brown tuber is rounded, 1.5–3.3 cm wide, and the attached roots are short and fleshy. Flowers open in spring and summer, or after rain inland.

Bulbine lily grows in conspicuous colonies in woodlands, grasslands and open forest.

USES: Aborigines harvested the bland starchy tubers.

GOLDEN STARS

Hypoxis

OTHER NAMES: Golden weather glass, star grass, nut lily, nut grass

FIELD NOTES: The yellow star flowers sprout singly or in irregular open clusters of two to six flowers, but not in tight spikes like bulbine lily. The very slender leaves and stems are often hairy. The single tiny tuber is egg-shaped or cylindrical. These are usually very delicate plants about 10–15 cm tall, found among grass. There are ten species.

USES: Aborigines ate the roasted tubers of some species (*H. pratensis, H. hygrometrica, H. nervosa*) though others are irritant and inedible.

C. spicatum

ARDA

Cartonema

FIELD NOTES: Flowers are yellow, with three petals, on hairy stalks, and open from late summer to winter. *C. spicatum* has buds crowded into a dense prickly head. *C. parviflorum* (illustrated page 195) has a taller stalk with widely-spaced buds, and does not look or feel prickly. Tubers are illustrated on page 20. Ardas grow in open forest in sandy or stony soil. *S. parviflorum* likes soggy soil, sometimes forming dense colonies. *S. spicatum* sprouts tubers only at the end of the wet season after its foliage wilts.

USES: Aborigines ate the cooked starchy tubers, which were once an important food.

GRASS POTATO

Curculigo ensifolia

FIELD NOTES: Leaves are very grass-like, held upright, 10-70 cm tall, narrowing at their base, folded like a fan in cross-section, and are often hairy. Tiny, yellow, star flowers with six petals sprout at ground level in summer and autumn. The taproot is long (to 16 cm) and cylindrical.

Grass potato grows on rocky knolls or in grassy flats in open forests, usually near streams or soaks. It is uncommon in the south. It resembles blady grass and is not easily spotted.

USES: Aborigines ate the slender, starchy taproots, raw or cooked. These taste sweet or bitter.

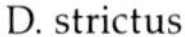
D. strictus

Chocolate Lilies

Dichopogon strictus, D. fimbriatus

Field Notes: The drooping flowers are pale mauve to dark purple, with crinkly, 5-14 mm long petals, easily identified by their chocolate or caramel smell, carried on a wiry stalk up to 1 m tall. Common chocolate lily (*D. strictus*) sprouts buds singly along the stalk; buds of nodding chocolate lily (*D. fimbriatus*) are in groups of two to six. Watery egg-shaped tubers, 1-3.5 cm long, sprout at the tips of the roots.

Chocolate lilies grow in colonies in woodlands and open forests among grass. Flowering is from August to February.

Uses: Aborigines ate the juicy, slightly bitter tubers.

Pink Swamp Lily

Murdannia graminea

Other Names: Grass lily, slug lily

Field Notes: Flowers are pink, lilac or lavender (rarely blue or white), with three rounded petals, in loose irregular sprays, and close in the early afternoon. Leaves are usually hairy, and their bases sheath the stalks, which are 15-30 cm tall. Six to 20 slender tubers lie clustered at the base of the plant. Flowers appear in summer and autumn.

Pink swamp lily prefers soggy sunlit ground near creeks and soaks, where it grows in colonies among grass.

Uses: Some Northern Territory tribes ate the very bland, fibrous tubers which are edible raw.

T. tuberosus

Common Fringed Lily

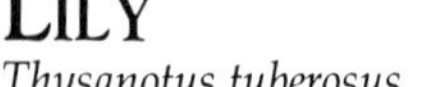

Thysanotus tuberosus

Other Name: Fringed violet

Field Notes: The mauve petals are fringed, 7-19 mm long; they open once during the morning, closing by early afternoon. A dozen or so whitish ellipsoid tubers, 1-2 cm long, are produced at the ends of the roots. Flowering is from spring to autumn.

Fringed lily grows scattered in open forests and woodlands. It is a well-known and much-loved wildflower, often featured on tea towels and place mats.

Uses: Aborigines ate the crisp juicy tubers. These are watery and almost flavourless.

Australia has another 46 species of fringed lily, found mainly in south-western Australia (where many are rare), but also in the south and north. About half the species produce tubers, and these were probably Aboriginal foods, although little is known about them. Three recorded foods are described on the following page.

T. baueri

Twining Fringed Lily

Thysanotus patersonii

Field Notes: This is a tiny, wiry, leafless creeper found twining among grass or over low branches in sunlit woodlands, heaths, and mallee. The slender mauve petals are 8-11 mm long. The flowers open in sprays during the morning from July to December; each flower opens only once. This is the only fringed lily that twines, apart from *T. manglesianus* in Western Australia.

Uses: Aborigines ate the watery tubers. In Western Australia the stems of this vine (or *T. manglesianus*) were rolled into a ball, baked in ash, ground to powder, and eaten with eucalypt gum.

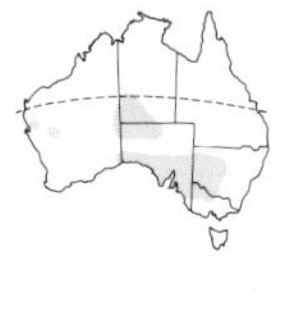

Desert Fringed Lilies

Thysanotus

Field Notes: (Illustrated previous page.) These lilies resemble the common fringed lily but the petals are much narrower. The leaves usually die away before flowering. The desert fringed lily (*T. exiliflorus*), found in South Australia, the Northern Territory and Western Australia, has rigid flowering stalks that branch more or less at right angles; the mallee fringed lily (*T. baueri*), found in southern South Australia, New South Wales and Victoria, produces its small flowers on arching upright stems.

Uses: Outback Aborigines ate the slender tubers. The desert Pintupi squeezed the tuber juice into the mouth when thirsty.

Spiranthes sinensis

Ground Orchids

family Orchidaceae

Field Notes: Ground orchids are small plants with up to six grass-like leaves and small flowers with six petals, at least one of which is lip-shaped. Australia has over 350 species.

All ground orchids produce tubers (see tuber gallery). These are pale and juicy, rounded, cylindrical or egg-shaped, one to eight in number. The potato orchid or cinnamon bells (*Gastrodia sesamoides*) has a tuber of sweet potato size, up to 15 cm long and 4 cm thick.

Ground orchids are found in most habitats, but especially in heaths and woodlands on sandy and stony soils.

Uses: Native orchids are often thought to be rare, but many Australian species are common, especially in the south where their starchy tubers were important Aboriginal foods. Botanist Beth Gott has counted 440 greenhood orchids in a single square metre of soil, yielding 880 tubers with a combined weight of only 126 grams. Tubers like these were roasted or eaten raw by Aborigines. They taste variously starchy, watery, bitter or sweet.

Orchid tubers were eaten by First Fleet convicts, and later by colonial lads, as noted by Joseph Maiden in 1898: "There is hardly a country boy who has not eaten so-called Yams, which are the tubers of numerous kinds of terrestrial or ground-growing orchids."

Epiphytic tree orchids such as the king orchid (*Dendrobium speciosum*) and "native arrowroot" (*Cymbidium canaliculatum*) lack tubers but have thickened stems which can be chewed and sucked for their starch. Colonists sometimes extracted this. Tasmanian Aborigines reportedly ate the leaves of ground orchids.

Eriochilus cucullatus

MURNONG

Microseris scapigera

OTHER NAMES: Yam daisy, native dandelion, *Microseris lanceolata*

FIELD NOTES: Murnong has yellow, dandelion-like flowers and a basal cluster of slender leaves, which are usually toothed (see leaf gallery). It is distinguished from dandelions and other weeds by its drooping buds, characteristic seeds, 7–10 mm long (see page 207), slender leaves, mostly 2-10 mm wide, and pale juicy taproot, 2-8 cm long, shaped like a radish or blunt carrot. The flowers open in spring and summer.

Murnong grows in grasslands, woodlands, open forests, and in Western Australia on saltpans. One alpine form in the Snowy Mountains has fibrous roots too thin to be worth eating. Grazing animals have exterminated it from pasturelands.

USES: Murnong tubers were favourite staple foods of Victorian Aborigines, gathered by the basketful from meadows and woods. According to one early report the tubers "were so abundant and so easily procured, that one might have collected in an hour, with a pointed stick, as many as would have served a family for the day". The tubers were baked in baskets or in holes in the ground "where they half melt down into a sweet, dark-coloured juice", one observer wrote in 1837. The tubers could be dug in any season, and are edible raw, though apt to taste bitter, especially in winter.

South Australian colonists imitated the Aborigines by eating murnong, and colonial botanist Baron von Mueller suggested it be cultivated as a vegetable in cold countries. It is one of Australia's tastiest staple foods, though low in protein and vitamins.

G. solanderi

CRANESBILLS
Geranium

OTHER NAME: Native geranium

FIELD NOTES: Cranesbills are small trailing plants with thick taproots and small, rounded, lobed leaves. The flowers are produced singly or in pairs; they have five white, pink or purplish petals 2-14 mm long. The seed pods are shaped like crane's bills, splitting characteristically when brown, as shown in the diagram.

Australia has at least ten cranesbill species found in woodlands and forests.

USES: The cooked taproots of cranesbills were reportedly eaten by Aborigines in Tasmania, Victoria, western New South Wales and Western Australia. Cranesbill roots I have tasted were dark red, intensely astringent and completely inedible. It seems likely that only a few species, or only the new season's pale taproots, are edible. In Western Australia, starving explorer George Grey ate cranesbill roots in 1839, recording in his journal: "Some of the men finding a species of geranium, with a root not unlike a very small and tough parsnip, we prepared and eat several messes of this plant."

The related native storksbill (*Pelargonium australe*) of temperate beaches and woodlands, also has an astringent red taproot reportedly eaten by Aborigines.

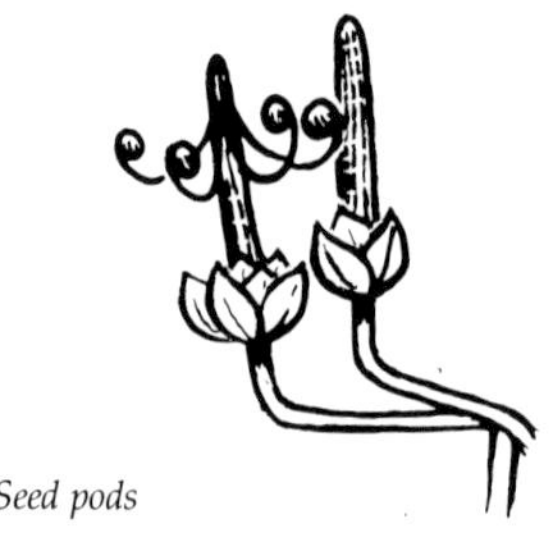

Seed pods

WILD PARSNIP

Trachymene incisa

OTHER NAME: Yam

FIELD NOTES: The pretty white flowerheads, produced mainly in summer, are 1-2 cm wide, on long wiry stems, 0.4-1.5 m tall. The lace-like leaves are 2-4 cm wide and deeply divided (see leaf gallery). The taproot resembles a small parsnip (see tuber gallery). The illustrated example is a very large old plant; most are smaller, with fewer flowerheads on more upright stems. They are inconspicuous apart from the flowers, which can be spotted from afar in the grassy understorey.

Wild parsnip grows in colonies on sandy infertile soils in coastal heaths and rocky outcrop country. It is patchily distributed but common where it occurs.

USES: Wild parsnip is one of Australia's tastiest wild foods. The thick juicy taproots are sweet and fragrant, and milder in flavour than their namesake. Edible raw or cooked, the young roots are especially succulent, becoming somewhat fibrous with age.

Wild parsnip is related to both the parsnip and carrot, and warrants cultivation as a vegetable, though it appears to be slow growing, reaching parsnip size only after several years.

There are no historical records of Aborigines eating wild parsnips, which only shows how incomplete our knowledge of Aboriginal diet must be. The plant is well known and much liked by country people.

NATIVE GOOSEBERRY

Physalis minima

FIELD NOTES: Leaves are soft, smooth (not furry), with smooth or jagged margins, 2.5–12 cm long. Cream flowers are followed by yellowish or purplish-grey berries inside papery husks which turn straw-coloured and fall to the ground when ripe. Native gooseberry grows about 20–50 cm tall and occurs on river banks, roadsides, fields and farms throughout the tropics of the world. It may not be native to Australia.

USES: Aborigines eat the fruits, which have a tangy cherry-tomato taste. Cape gooseberry (*P. peruviana*), a taller weed with furry leaves, also has edible fruits, which were widely eaten by colonists.

AMULLA

Myoporum debile

OTHER NAME: Winter apple

FIELD NOTES: Leaves are slender, 3-12 cm long, on very short stalks, usually with a few serrations at the base. The ripe fruits are purplish or pink, sometimes blotched, 6-8 mm long and slightly flattened, contain one flat seed and have five triangular leaf-like bracts at the base. They ripen in summer and autumn. Amulla forms mats of creeping foliage in sunny open woodland. It is uncommon and easily overlooked.

USES: Aborigines ate the small fruits, which taste salty-sweet when completely ripe. Most of the fruit consists of the large seed.

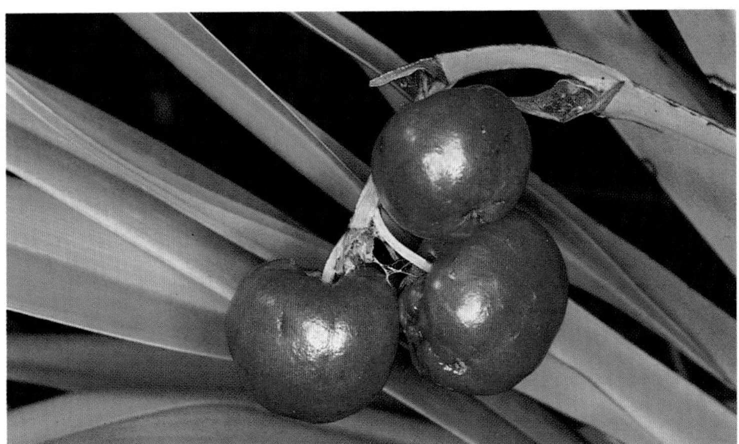

D. congesta

FLAX LILIES

Dianella

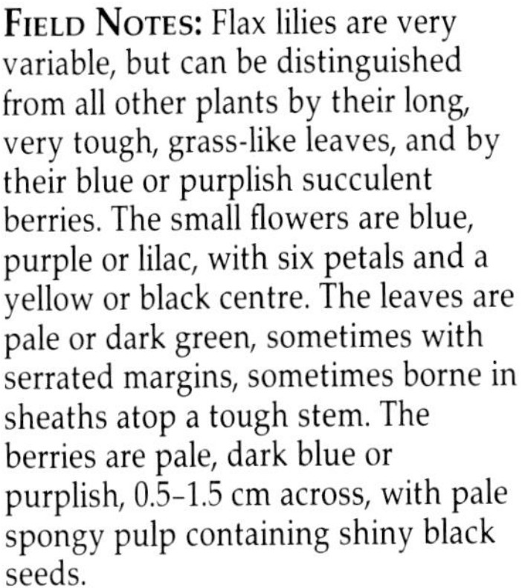

FIELD NOTES: Flax lilies are very variable, but can be distinguished from all other plants by their long, very tough, grass-like leaves, and by their blue or purplish succulent berries. The small flowers are blue, purple or lilac, with six petals and a yellow or black centre. The leaves are pale or dark green, sometimes with serrated margins, sometimes borne in sheaths atop a tough stem. The berries are pale, dark blue or purplish, 0.5–1.5 cm across, with pale spongy pulp containing shiny black seeds.

The beach flax lily (*D. congesta*) grows in swards on coastal dunes between Rockhampton and Wollongong, and carries its large fruits in tight bunches on an arching stem which is barely taller than the leaves. Most other flax lilies produce their fruits in open sprays atop an upright stem, which can be up to 1.5 m tall and is held well above the foliage.

Australia has 15 species of flax lily, found in moist forests, dry woodlands, rainforests and coastal dunes.

USES: Aborigines used the tough leaves of flax lilies to weave dillies and baskets, and no doubt ate the berries as well, although no record of this survives. I have sampled many species from eastern Australia, and all appear to be edible apart from a couple of southern forest forms with irritating berries, including the big-fruited *D. tasmanica*. Any that taste good are safe to eat in small amounts. The beach flax lily has tasty berries but most kinds are very insipid. Edible species include *D. caerulea*, *D. longifolia* (formerly *D. laevis*), *D. revoluta*, *D. pavopennacea* and *D. bambusifolia*.

BLADY GRASS

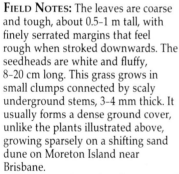

Imperata cylindrica

FIELD NOTES: The leaves are coarse and tough, about 0.5-1 m tall, with finely serrated margins that feel rough when stroked downwards. The seedheads are white and fluffy, 8-20 cm long. This grass grows in small clumps connected by scaly underground stems, 3-4 mm thick. It usually forms a dense ground cover, unlike the plants illustrated above, growing sparsely on a shifting sand dune on Moreton Island near Brisbane.

Blady grass is an abundant grass of coastal forests, woodlands, beaches and pastures, especially on impoverished soils. It occurs throughout the warmer regions of the world.

USES: Blady grass is related to sugarcane (which is also a grass), and the two have been interbred experimentally in India. Both have stems rich in sugar, although blady grass stems grow underground, where they are insulated from fires and grazing. Blady grass thus thrives in areas ravaged by burning and overgrazing. It is the dominant plant of abandoned slash-and-burn clearings in Asia.

In eastern Asia the sugary stems of this grass were fermented into alcohol, but there is little record of their use by Aborigines. At Belyuen near Darwin, the stems were given to young children to chew upon while the mothers dug yams. On Stradbroke Island, near Brisbane, children sucked the roots like sugarcane. The stems are too insubstantial to serve as an adult food, and they are not regarded as edible by most Aboriginal communities.

Aborigines used blady grass leaves to thatch huts and weave dillies.

BRACKEN

Pteridium esculentum

OTHER NAME: *P. aquilinum* var. *esculentum*

FIELD NOTES: The fronds are dark green, glossy and hard, on stiff stalks joined to hairy rhizomes (underground root-like stems). The thickened rhizomes contain slimy white starch. Bracken grows about 0.7–1.5 m tall, sometimes much taller. It can be confused with the soft-leaved, pale green false bracken.

Bracken grows in forests, heaths and paddocks, often as the dominant ground cover.

USES: In south-eastern Australia, from southern Queensland to Tasmania, bracken starch was an important Aboriginal food. In southern New South Wales, for example, the roasted rhizomes were staple foods, eaten in huge quantities when fish were scarce. Around Sydney fern starch was flavoured with crushed ants. The stems beneath emerging bracken shoots appear to contain the most starch, which is slimy and tasteless. Older rhizomes are fibrous and insubstantial. In Tasmania, Aborigines also ate the slimy young shoots emerging from the ground.

Overseas, bracken was a staple food of the New Zealand Maori, and European bracken was an emergency food during the Irish potato famine.

Boiled fern shoots, called croziers or fiddles, are often cited in survival manuals as emergency foods. The fiddles of bracken are inclined to taste bitter, and, as the fronds are known to contain toxins, including carcinogens, they should not be eaten in quantity. In Japan a high incidence of bowel cancer has been blamed on consumption of bracken fiddles.

DYSENTERY BUSH

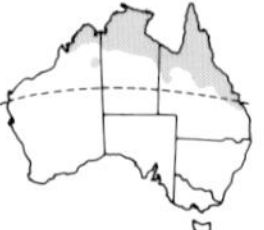

Grewia retusifolia

OTHER NAMES: Emu-berry, dogs nuts, turkey bush, diddle diddle

FIELD NOTES: This is a small twiggy plant, up to a metre tall but usually smaller, with coarse, heavily veined, toothed leaves, pale and fuzzy on the undersides. The leaves are slender, 6-12 cm long, alternate, dull green to slightly glossy, and vary in shape from slender (as illustrated) to a very broad (when growing in shade). Starry white flowers are followed in late autumn and winter by small reddish-brown fruits, with two to four lobes, each containing a hard seed surrounded by sweet dry flesh. The fruits are shiny, hairy, and measure about 1 cm across.

Dysentery bush is common in open woodland on sandy or rocky soils, but is easily overlooked.

USES: Despite their small size, the sweet fruits are a very popular snack food of Aborigines. They are rich in thiamine, yielding about 1000 micrograms per 100 grams, and contain some vitamin C. Explorer Ludwig Leichhardt found they had a "very agreeable taste"and made an excellent drink by boiling them in water.

Dysentery bush is one of the most important Aboriginal remedies in northern Australia. The leaves are chewed, or an infusion is drunk, for diarrhoea and upset stomach, and a preparation of the roots is applied to boils, sores, and scabies.

Related species of *Grewia* are also edible. *G. latifolia,* found in northern and eastern Australia as far south as Glen Innes, is a very similar plant to dysentery bush, with similar fruits, but with broader, oval leaves. *G. multiflora,* a large shrub very common around Darwin, has broader leaves, yellow flowers, and bluish-black spherical fruits that taste tangy. *G. breviflora* of northern Australia is also edible.

Polynesian Arrowroot

Tacca leontopetaloides

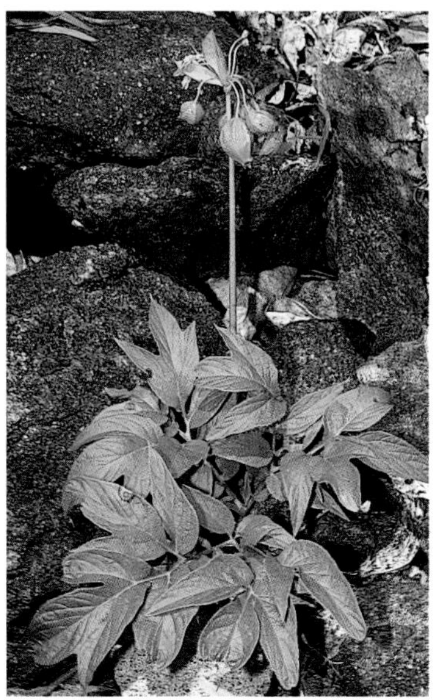

Field Notes: Polynesian arrowroot produces only a single very large leaf, but this is divided into three very segmented sections and thus resembles a spray of leaves. Greenish or yellowish flowers and fruits are produced in a cluster atop a thick green stem which is 0.5–2.2 m tall. The flowers are surrounded by leafy bracts and very long, hair-like threads. The dangling segmented fruits contain many seeds and turn pale orange when ripe. The plant sprouts from one to two smooth, potato-like tubers.

Polynesian arrowroot is an unmistakable plant, often found growing behind sandy beaches, on rocky slopes, or scattered in open forests and monsoon rainforests. It ranges from Australia to the South Pacific and southern Asia.

Uses: At the end of the tropical wet season the foliage of Polynesian arrowroot dies away, and the plant survives the dry winter months as a big, starch-filled, underground tuber. This was eaten by Aborigines after careful preparation to leach out a very irritating toxin. The tuber was usually grated or pounded, soaked in fresh water, and the resultant bland mush was cooked. Each tribe had its own method of preparation, as recounted in detail by ethnographer Walter Roth. On the islands of Torres Strait the grated soaked tubers were baked in coconut milk. In Asia and the South Pacific the starch was an important food. I have eaten it raw after simply grating a tuber and soaking it for 36 hours in mosquito netting suspended in a stream. It tasted like raw potato. The succulent fruits are also edible.

LONG-LEAF MAT-RUSH

Lomandra longifolia

OTHER NAMES: Spiny-headed mat-rush, sag

FIELD NOTES: This rush is identified by its strap-shaped leaves about 1 cm wide with ragged tips, and by its spiky flower and seed spikes. The flowers are tiny, yellowish, and heavily scented. It grows in thick, untidy clumps 1-2 m tall.

Mat-rush is very common along streams and amongst granite outcrops, but also grows in low-lying woodlands, heaths, alpine meadows, and on sea cliffs and dunes. It is widely cultivated.

USES: The raw leaf bases of mat-rush provide a refreshing snack for the bush rambler. Tufts of the leaves are pulled from the clump and the white inner bases are chewed. These taste like fresh green peas.

Many kinds of sedges and rushes have white edible leaf bases and it is worth tasting these at random. For example, colonists in southern Australia chewed the bases of the coast sword-rush (*Lepidosperma gladiatum*), a common sedge of southern seashore dunes with thick, sharp-pointed leaf blades and brown or black seedheads. Saw sedges (*Gahnia*), plants with spiny-edged leaves found in swamps and swampy heathlands, can also be harvested in this way.

Aborigines in Victoria reportedly ate the "sweet flowers of several species of *Xerotes* [*Lomandra*]", including possibly this species. Other kinds of *Lomandra* in Victoria are much smaller woodland plants with spineless sprays of flowers.

The tough leaves of the long-leaf mat-rush were split into strips and woven by Aborigines into sturdy dillies and mats.

Flame Lily

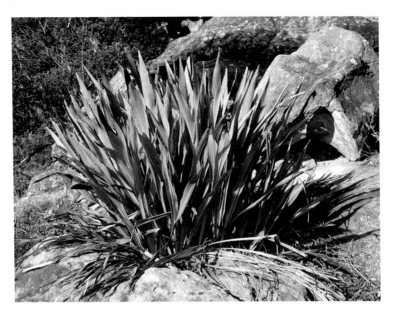

Doryathes excelsa

Other Names: Gymea lily, giant lily, spear lily, Illawarra lily

Field Notes: The huge sword-shaped leaves are a metre or so long and up to 10 cm wide. In spring and early summer the plants send up tall flowering spikes, 2–4 m tall, which are topped by clusters of enormous, dull red, trumpet flowers, which attract nectar-feeding birds.

Flame lily grows only on sandstone ridges in the Hawkesbury sandstone belt, between Newcastle and Wollongong. It is common at Waterfall. It is sometimes cultivated.

Uses: Quaker missionary James Backhouse, while travelling near Newcastle in 1836, recorded the Aboriginal use of this plant:

> Many of the open places in the forest, abounded with Gigantic lily; the flower stems of which rise from 10 to 20 feet high. These stems are roasted, and eaten by the Aborigines, who cut them for this purpose, when they are about a foot and a half high, and thicker than a man's arm. The Blacks also roast the roots, and make them into a sort of a cake, which they eat cold.

Colonial botanist Joseph Maiden noted that "the leaves are a mass of fibre of great strength, suitable for brush-making, matting, etc."

D. bulbifera, *showing bulbils*

YAMS
Dioscorea

FIELD NOTES: Yam leaves are typically heart-shaped with veins that radiate from the top of the leaf where the stalk is attached. Stems are thin and wiry without milky sap. The minute flowers hang in strings. The seed pods have three flat wings and hang in clusters. Yams resemble other jungle vines and the pattern of leaf veins should be closely compared.

Identification of Australian yams is somewhat confused, but the following species are recognised.

The round or hairy yam (*D. bulbifera,* formerly *D. sativa*) has leaves with seven to 11 longitudinal veins, and with strongly marked cross-veins. The seed pods have narrow wings, 2-3 cm long, 6-7 mm wide, and the stems produce bulbils (tiny aerial yams) late in the Wet season. The tuber is rounded and hairy with yellowish flesh. This vine is common in monsoon rainforests and open forests north of Mackay.

The long or pencil yam (*Dioscorea transversa*) has shiny leaves with five or seven prominent veins and without conspicuous cross-veins. The seed pods are rounded and the yam is long and usually slender with white or pale grey flesh. There are two forms of long yam: one found in rainforest clearings and wet forests in eastern Australia which does not bear bulbils (described and illustrated page 61); and a northern form found in sandy soil in coastal open forests, often behind beaches. This has enormous tubers and sprouts small bulbils.

Warrine, or warrien (*D. hastifolia*, see leaf gallery) has very slender leaves, only 2-20 mm wide, and rounded pods. It grows in open forests in south-western Australia.

As well as the native yams, several Asian species now run wild in northern Australia. The greater yam (*Dioscorea alata*), found widely in the north, has four flaps or ridges running along its stems. It is a very lush leafy vine with an enormous tuber. The prickly yam (*D. nummularia*) found in Torres Strait has prickly stems, and the five leaf yam (*D. pentaphylla*) of Thursday and Hammond Islands has five-lobed leaves.

D. transversa

Uses: The vines of genus *Dioscorea* are the true yams. Though often assumed to be typical Aboriginal tucker, yams are confined to coastal areas beyond reach of most tribes; however, wherever they occurred they were staple foods.

The round yam has "cheeky" (poisonous) tubers which need to be grated and soaked (as well as cooked) before eating. On some north Queensland islands the grated tubers were boiled in baler shells. Aborigines applied simple gardening techniques by replanting the tops after removing the yams, even planting the yams on offshore islands for future needs. The round yam occurs widely in Africa and Asia and may be a very early human introduction into Australia.

The warrine and the exotic yams can be eaten after roasting. The long yam, an extremely popular food, can be eaten raw or roasted.

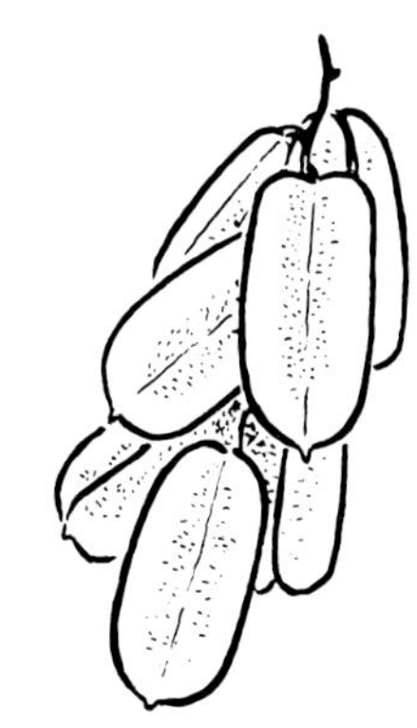

D. bulbifera *seed pods*

Flowers and fruit of E. latifolius

WOMBAT BERRY

Eustrephus latifolius

OTHER NAMES: Orange vine, black-fellow's vine

FIELD NOTES: Wombat berry is a wiry vine which trails over low branches or across the ground. The soft, bright green leaves are 3-11 cm long, 0.5-4 cm wide, paler on the undersides, slender (as illustrated) or broad, and finely lined with parallel longitudinal veins. Flowers are white or lavender-pink, with six "petals", three of them fringed. The 1-1.5 cm wide berries ripen from green to orange, splitting to display shiny black seeds.

Wombat berry is a common vine of open forests, extending into dry woodlands and rainforest. It ranges from east Gippsland to New Guinea and New Caledonia. It should not be confused with the very similar scrambling lily (*Geitonoplesium cymosum*, see below), which has shiny black (inedible) berries and unfringed white petals.

USES: The burst berries of wombat berry contain a tiny amount of crisp white pulp, which was eaten by Aborigines. The roots swell into small earthen-coloured tubers, 1-3 cm long, which taste sweet and juicy, unless dry weather has shrivelled them and made them bitter. Aborigines also ate these, although they are not easy to dig from the hard soil in which they usually grow. The nineteenth century botanist Baron von Mueller, a man of some imagination, envisaged them as a food crop "probably capable of enlargement through culture".

The related scrambling lily of moist forests and creek banks resembles wombat berry, except its berries are black. It has shoots which colonists compared to asparagus, for they may be boiled and eaten.

WILD MUNG BEAN

Vigna radiata

OTHER NAMES: Finger bean, green gram, [*Phaseolus mungo*]

FIELD NOTES: The mung bean is a creeper with leaves in threes, bristly stems, and greenish-yellow pea flowers about 1-1.5 cm long. The innermost "petal" (the keel) of each flower curves asymmetrically and contains a small pocket. The seed pods are green, bristly, about 4-6 cm long, turning black and splitting when ripe to catapult the seeds.

The mung bean is a creeper of grassy open forests and roadsides.

USES: Mung bean plants found growing wild in Australia are usually dismissed as escapees from farms. But dried specimens held by the British Museum, collected by Joseph Banks at Endeavour River in 1770, show the mung bean to be a native plant in Australia. Its natural range is now known to extend from India to China and south through Indonesia to Australia. In Asia the seeds and seed sprouts are cooked in a variety of dishes, and the young green pods are sometimes eaten.

Aborigines in north Queensland harvested mung bean plants, not for their seeds, but for their slender taproots, which were roasted and eaten.

The grey-black seeds of Australian mung beans are only one quarter the size of cultivated forms, and though very hard, are still good to eat. The green pods can also be eaten like raw beans.

The Australian and Asian forms are being interbred to create commercial hybrids suitable for Australian conditions. The market for sprouts in Australia is small, but the seeds bring high prices in Asia.

B. cymosa

B. scandens

APPLE-BERRIES
Billardiera

OTHER NAME: Dumplings

FIELD NOTES: Apple-berries are small twining creepers with slender leaves and small, sausage-shaped berries containing many tiny seeds.

Common apple-berry (*B. scandens*) has greenish or pale yellow tubular flowers, and greenish, usually hairy fruits, 1-2.5 cm long, of variable thickness, borne singly. It grows in shady eucalypt forests, rainforests and heaths from southern Queensland to Mount Gambier and Tasmania.

Sweet apple-berry (*B. cymosa*) has pale, bluish, greenish or cream flowers, and 1-1.5 cm berries in clusters. It grows in woodlands, mallees and coastal heaths in western Victoria and South Australia.

Purple apple-berry (*B. longiflora*) has greenish-yellow flowers and shiny purple fruits. It grows in cool moist forests from southern New South Wales to Tasmania.

Australia has another 20 or so *Billardiera*, some of which are edible.

USES: Common apple-berry grows freely around Port Jackson and was one of the first Australian fruits to be eaten by Europeans. The botanist Joseph Maiden recorded in 1898 that "it has been eaten by children ever since the foundation of the colony, and is one of the earliest known food-plants of the blacks". Aborigines ate not only the ripe fruits, which taste like kiwi fruit, but also roasted and ate the unripe fruit. The fruits ripen only after they fall to the ground.

Sweet apple-berry fruits have a delightful aniseed flavour; purple apple-berry fruits are aromatic and mealy.

The similar-looking bluebell creeper (*Sollya heterophylla*), found in south-western Australia, also has edible fruit.

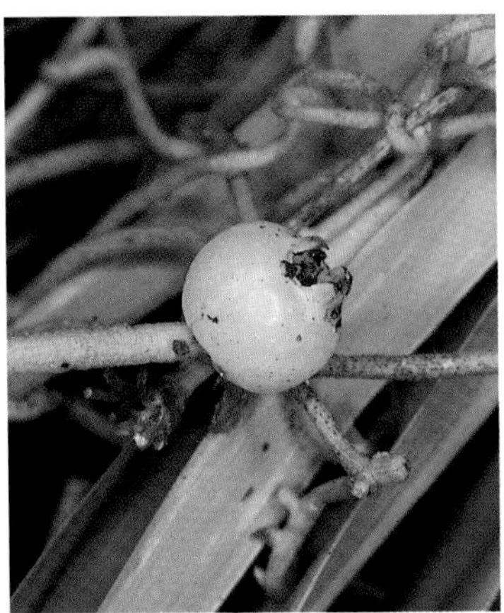

C. filiformis

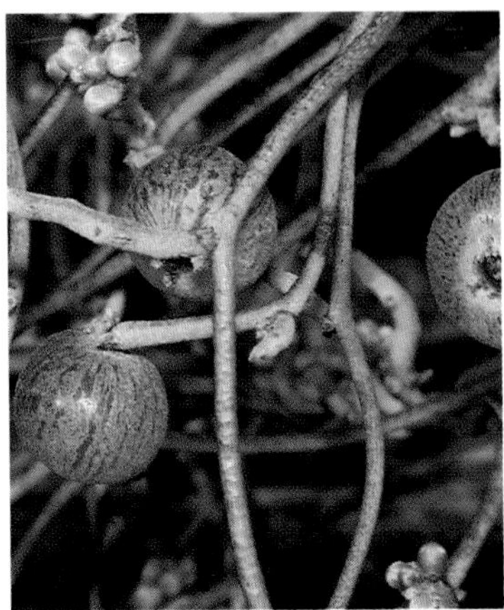

C. pubescens

Devil's Twines

Cassytha

Other Names: Dodder laurel, devil's guts, black-fellow's twine

Field Notes: Devil's twines are unmistakable plants, easily recognised by their twining, wiry, leafless stems and succulent fruits. They form untidy wiry mats across the crowns of shrubs. Dodder (*Cuscuta*), an introduced weed of farms and vacant allotments, is a similar-looking plant with leafless twining stems, but it does not produce juicy fruits.

Devil's twines are parasitic plants. They penetrate their host's stems and draw away nutrients, though they also produce chlorophyll and can photosynthesise. After germinating and attaching to a host, they lose all contact with the ground.

Australia's 14 species are plants of infertile sandy and rocky soils. Common in heaths, they sometimes form dense tangles on beach dunes. The stems are greenish, orange-yellow or purplish-red, 1-2 mm thick. The fruits are 4–15 mm long, rounded or oval, sometimes ridged, smooth or hairy, any colour except blue, and contain one round stone.

The most wide-ranging species is *C. filiformis*, easily recognised by its white, pearl-like fruits. It grows in northern and eastern Australia, especially on beaches, as far south as northern New South Wales.

Uses: The small fruits of devil's twines are edible. Most are resinous, sticky, and not very tasty. They served as snack foods of Aborigines. Velvety devil's twine (*C. pubescens*) was eaten by colonial children last century. Aborigines sometimes used the stems as twine.

Lysiana exocarpi

Viscum articulatum

MISTLETOES

families Loranthaceae, Viscaceae

FIELD NOTES: These parasitic shrubs drape from the branches of shrubs and trees. Australia has more than 80 species in two families, and they vary greatly in appearance. Leaves can be broad, slender or absent, and the fruits are round, oval or pear-shaped, 4-15 mm long and coloured yellow, red, pink, white or black.

Despite the variation, mistletoes are easy to recognise. Their foliage usually contrasts with the host, and the point of attachment is obvious. Leaves are rather thick and brittle, the stems are brittle, and the fruit pulp is always very sticky, with one seed, usually attached to a long sticky strand (see *Viscum articulatum* illustration).

Mistletoes grow in all habitats—from mangroves and jungles to deserts—on a spectrum of shrubs and trees. Some parasitise each other. The largest genera are *Amyema* (36 species) and *Lysiana* (eight species). (See also under Mulga, in the Arid Zone section, page 181.)

USES: Mistletoes could not survive without the mistletoe bird, which eats their fruits and voids the sticky seeds on live twigs, where the plants germinate. The bird's stomach is a special tube for processing mistletoe fruits, which pass through in 25 minutes.

Aborigines snacked upon the fruits of many kinds of mistletoes, especially *Amyema* and *Lysiana* species, though not all were used. The pulp is pleasantly sweet, but very sticky.

On Stradbroke Island near Brisbane the Aborigines made chewing gum by chewing the half-ripe fruits.

Pink-flowered Native Raspberry

Rubus parvifolius

Other Name: Small-leaf raspberry

Field Notes: This is a small and very scrawny shrub or creeper with prickly arching canes. The leaf segments are tough and crinkly, with white undersides; they sprout in groups of three or five. Pretty pink flowers are followed by raspberries about 1 cm wide, which ripen in late summer and autumn.

This raspberry grows in open forests, especially on rocky hillsides and on sand behind beaches. It is also found naturally in Japan, where the fruits are made into wine.

Uses: Unlike Australia's "native cherries", "wild oranges", "crab apples" and the like, which bear no relation to their namesake, the raspberries of eastern Australia are true raspberries, closely related to the species in Europe and America. They were widely gathered by colonists and made into jams and pies. Some of the native raspberries are insipid, but this species has an excellent flavour, and was probably the one alluded to by surgeon P. Cunningham in 1827 when he wrote: "Of native fruits we possess raspberries equal in flavour and not otherwise distinguishable from the English." He said they made "an excellent preserve" when mixed with native currants. The tangy fruits were also, of course, an Aboriginal food.

Australia has another six kinds of raspberry, five found near rainforest (see pages 66-67) and one in the mountains of Tasmania. This is the mountain or alpine raspberry (*R. gunnianus*), a small creeping herb with small, deeply serrated leaves and red berries consisting of only a few large shiny red segments.

HEATHS

family Epacridaceae

The fruit-bearing heaths of southern and eastern Australia show how little we know about Aboriginal diet. There are dozens of species with sweet edible pulp yet hardly any are recorded as Aboriginal foods. Early observers no doubt overlooked their use because of their small size—they were only snack foods.

At least nine heath genera have species with edible fruit. The total number of edible species is unknown; fruits of some species have yet to be sighted, much less tasted. The heaths illustrated here show a variety of forms, and hopefully convey the essential "heathness" of the group. It is safe to sample any heath fruit, and any that tastes good will be safe to eat.

Heaths are twiggy shrubs, rarely small trees, adapted to sandy and rocky infertile soils. Some alpine and heathland species grow as "sub-shrubs" forming ground-hugging mats of dense foliage. Heaths often dominate the habitats called "heaths" but also grow in forest understoreys, on mountain slopes, alpine meadows, and behind beaches. Most species are found in south-western and south-eastern Australia; only a few bearded heaths extend as far north as Cape York.

Heath leaves are small, usually 1-2 cm long, slender and hard, often rigid and spiny-tipped, without veins on the upper surface, and with pale green or whitish undersides showing parallel, closely spaced longitudinal "veins". Flowers are mostly either tiny and white (insect pollinated) or long, tubular and coloured (attracting birds). The five petals are often furry. The twigs are brittle.

The fruits are small, usually less than 1 cm long, oval or round, coloured red, orange, purple, yellow, white or green, and always contain one hard stone—they are often tipped by a stalk-like style, and are sometimes concealed beneath the foliage. The soft pulp is pleasantly sweet, and readily acceptable to the palate.

Nearly all the heaths appear to have edible fruits. The only exception I know is the tree heath (*Trochocarpa laurina*) of eastern rainforest margins, which has big broad leaves, 4-8 cm long, and blue-black or yellowish bitter fruits.

Heath fruits are eaten by emus, small birds and rats. Some varieties were popular with early colonists, especially native cranberry (*Lissanthe sapida*) eaten around Sydney, and native currants (*Acrotriche depressa*) and spreading brachyloma (*Brachyloma depressum*), harvested for jams and jellies in South Australia and Victoria respectively.

The following genera are known to include edible species: *Acrotriche, Astroloma, Brachyloma, Cyathodes, Leucopogon, Lissanthe, Melichrus, Monotoca, Pentachondra* and *Styphelia*.

Aborigines also ate the tiny green flowers of one southern heath (*Acrotriche serrulata*, which also has edible green fruit), and probably sucked the nectar of honey pots (*Melichrus procumbens*).

S. viridis

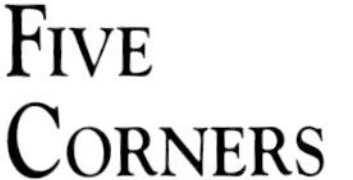

Five Corners

Styphelia triflora, S. viridis

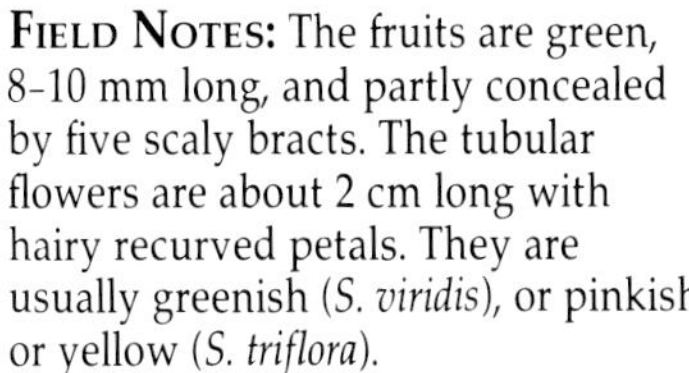

Field Notes: The fruits are green, 8-10 mm long, and partly concealed by five scaly bracts. The tubular flowers are about 2 cm long with hairy recurved petals. They are usually greenish (*S. viridis*), or pinkish or yellow (*S. triflora*).

These heaths are scrawny shrubs, 1-2 m tall, found in heathlands and woodlands on sandy or stony infertile soil. *S. viridis* also grows on old dunes near the sea.

Uses: Aborigines ate the sweet tasty fruits. The very similar golden heath (*S. adscendens*) of Victoria, Tasmania and southern South Australia, is also edible. Its fluffy flowers are greenish-yellow.

Prickly Broom Heath

Monotoca scoparia

Field Notes: The lemon or white fruits are 2-3 mm wide, and the leaves are dark green, rigid, convex, 5-23 mm long, with spiny tips. This is a low shrub or (in southern Queensland) a small tree, found in seashore heaths and in woodlands on sandy infertile soils. It resembles coast beard heath (see page 37) but has fruits on shorter stalks, and leaves with whitish undersides.

Uses: The fruits taste sweet. The similar tree broom heath (*M. elliptica*) of seashore scrubs and forests has orange-red, edible fruits.

MOUNTAIN BEARD HEATH

Leucopogon suaveolens

OTHER NAME: *L. hookeri*

FIELD NOTES: This is a small shrub of alpine heaths and mountain forests above 1000 m altitude, found as far north as the New England Tableland. The tiny, white, furry flowers are followed by 4-6 mm wide, red fruits, which ripen in summer and autumn. The leaves are 5-14 mm long and 1-4 mm wide, with curled margins and leaf tips that are not tapered.

USES: The ripe fruits taste sweet. The similar snow beard heath (*L. montanus*) of high alpine meadows also has white, furry, flowers and red edible fruits. Its leaves are shorter (3-7 mm long) and flat. Nearly all kinds of red alpine fruit are edible.

LANCE BEARD HEATH

Leucopogon lanceolatus

FIELD NOTES: The slender tapered leaves are 1.5-7 cm long and the scarlet fruits are only 2-3 mm wide, on spikes 1.5-4 cm tall. Sprays of pink buds appear in autumn, followed by tiny, furry, white flowers in spring, and shiny fruits in summer.

This 1-3 m tall shrub grows in mountain forests, on rocky slopes, and occasionally in coastal heaths. It is very common in the upper Blue Mountains.

USES: The sweet watery fruits were eaten by colonists, and no doubt by Aborigines. Naturalist George Bennett in 1860 described them as having "a pleasant subacid taste".

CARPET HEATH

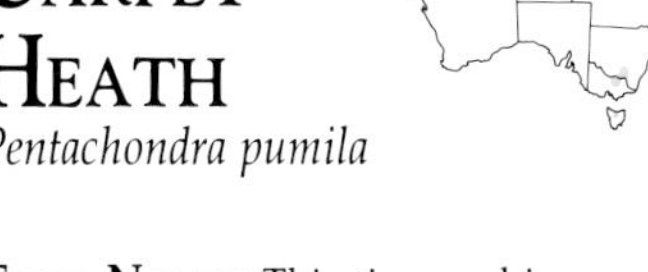

Pentachondra pumila

FIELD NOTES: This tiny cushion plant of alpine meadows grows only 5-15 cm tall. Its furry, white flowers and 5-8 mm wide, crimson fruits often sprout simultaneously. The stiff leaves are 3-6 mm long.

USES: Fruits are sweet and watery.

TALL GROUNDBERRY

Acrotriche aggregata

FIELD NOTES: The shiny, bright red, sour fruits sprout along the twigs beneath the foliage, and ripen almost any time of year. This is a shrub of sandy infertile soils on hills and mountains.

USES: The fruits taste tangy.

Cranberry Heath

Astroloma humifusum

Other Name: Native cranberry

Field Notes: This small, ground-hugging shrub grows only 10-30 cm tall. Leaves are stiff, narrow, spine-tipped, 5-12 mm long, with hairy edges. Scarlet tubular flowers are followed by greenish fruits, 7-11 mm wide, often with purplish spots or stripes, which ripen from June to November. Fruits are often concealed within the foliage. This heath grows in heathlands and woodlands on sandy and stony soils.

Uses: Botanist Joseph Maiden wrote in 1889: "Fruits of these dwarf shrubs are much appreciated by school-boys and aboriginals." They taste like apples.

Flame Heath

Astroloma conostephioides

Field Notes: Flame heath is a prickly, twiggy shrub, 0.3-2 m tall, with narrow, dark green, spine-tipped leaves, 1-2.5 cm long. Spectacular scarlet tubular flowers are followed in spring and summer by 1 cm whitish fruits completely enclosed inside reddish-brown papery scales.

Flame heath grows in heaths, mallees and forests. It is common in the Grampians, in the Mount Lofty Ranges, and on Kangaroo Island.

Uses: The fruits are sweet and sticky. They are avidly eaten by emus, which also browse upon the flowers.

S. linearifolium

KANGAROO APPLES

Solanum

OTHER NAMES: See below

FIELD NOTES: Kangaroo apples are weedy, soft-leaved shrubs, mostly 1.5-2.5 m tall, with long, drooping, usually dark green leaves and shiny egg-shaped berries, 1.5-2.5 cm long, carried on long drooping stalks, containing numerous tiny, pale, flattened seeds surrounded by moist pulp. The flowers have five purplish joined petals and yellow centres. There are four species, one found mainly in rainforest (see page 68).

The southern kangaroo apple (*S. laciniatum*) produces orange or yellow berries in bunches, and has large leaves, 9-38 cm long, which are lance-shaped or often lobed, like those of the rainforest kangaroo apple, which it closely resembles (see page 68). This weedy shrub grows in southern Australia (and New Zealand), north to Dorrigo, New South Wales, in coastal dunes, damp forest valleys and on rocky slopes.

The mountain kangaroo apple (*S. linearifolium*), has yellowish berries with purplish markings and very slender leaves mostly 5-10 cm long, 0.5-1 cm wide. It grows on hillsides and mountains in Gippsland, southern New South Wales, and the Australian Capital Territory.

Gunyang (*S. vescum*) has big, ivory green berries and large, broad, lance-shaped or lobed leaves, 5-50 cm long. It grows on coastal sands, in forests, woodlands and rainforest clearings, in Tasmania, and from Melbourne to Gympie, Queensland.

USES: Aborigines ate the soft berries, which have a sickly sweet taste. The fruits of the southern and mountain kangaroo apples signal their ripeness by bursting.

P. virgata

Geebungs

Persoonia

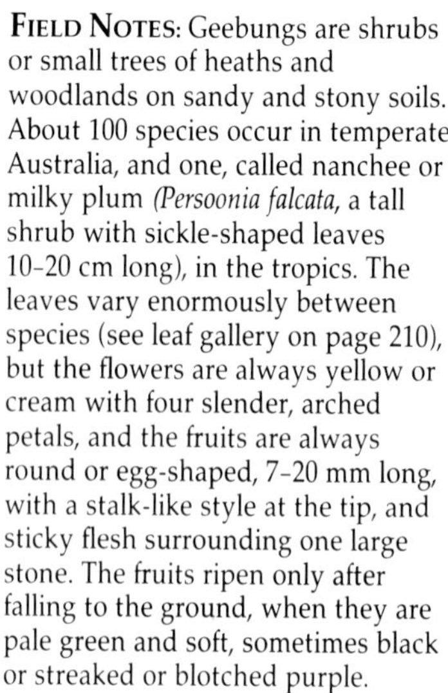

Field Notes: Geebungs are shrubs or small trees of heaths and woodlands on sandy and stony soils. About 100 species occur in temperate Australia, and one, called nanchee or milky plum (*Persoonia falcata*, a tall shrub with sickle-shaped leaves 10-20 cm long), in the tropics. The leaves vary enormously between species (see leaf gallery on page 210), but the flowers are always yellow or cream with four slender, arched petals, and the fruits are always round or egg-shaped, 7-20 mm long, with a stalk-like style at the tip, and sticky flesh surrounding one large stone. The fruits ripen only after falling to the ground, when they are pale green and soft, sometimes black or streaked or blotched purple.

Uses: Geebungs are a favoured food of kangaroos, rats, emus, birds, and nowadays, feral pigs. The sweet pulp can be nibbled as a snack, although it is fibrous, sticks fast to the large stone, and tastes like sweet cotton wool.

Geebungs were popular Aboriginal foods, and settler Constance Petrie told how the fruits were "greatly relished" by the Brisbane tribes: "The natives got dillies full of these in the right season. They swallowed the pulp and the stone, which they squeezed from the skin with their fingers."

Geebungs were eaten by colonial lads and sometimes by their elders. R. M. Praed wrote in 1885: "We gathered the wild raspberries, and mingling them with geebongs and scrub berries, set forth a dessert."

In the Kimberleys, nanchee fruits are sun-dried by Aborigines, roasted in ashes, hammered to crush the seeds, and stored in paperbark for future eating.

SOUR CURRANT-BUSH

Leptomeria acida

OTHER NAME: Native currant

FIELD NOTES: This leafless, broom-like shrub, 1.5–3 m tall, has foliage consisting of many slender, finely ridged green stems. Minute brownish flowers are followed by sour green fruits, sometimes with pale spots, 4–8 mm long, which contain one seed. They ripen from October to May.

Currant-bush grows in heaths and rocky woodlands on sandy infertile soils.

USES: Scurvy was the scourge of the first convict colony at Port Jackson and various wild foods were tried as cures. The best results came from "a kind of shrub in this country resembling the common broom", the Surgeon-General John White wrote in 1788, obviously referring to currant-bush. He considered it "a good anti-scorbutic [scurvy cure]; but I am sorry to add, that the quantity to be met with is far from sufficient to remove the scurvy".

Scurvy is caused by lack of vitamin C. Currant-bush fruits contain only moderate amounts of this vitamin—about 21 milligrams per 100 grams, compared to 50 milligrams in oranges. A cupful contains enough vitamin C for one person's daily needs, but four cups would be needed to cure scurvy. These amounts were probably unavailable.

Aborigines ate the "currants", and the missionary James Backhouse said the women carried large amounts in "calabashes".

Aborigines ate the fruits of related currant-bushes, including *L. drupacea* in Tasmania, and the leafless currant-bush (*L. aphylla*, illustrated page 197) of open forests and mallees in western Victoria and southern South Australia. It has spiny stem tips.

Native Cherry

Exocarpos cupressiformis

Other Names: Wild cherry, cherry ballart, forest cherry

Field Notes: This is a small tree with a dense crown of drooping leafless stems, resembling a she-oak or cypress, but with a more yellowish-green foliage. Tiny, insignificant greenish flowers are followed by a hard seed attached to a succulent red fruit, 4-6 mm long, ripening mainly in summer and autumn.

Native cherry is a root parasite that often suckers from its roots to form clumps of seemingly separate shrubs or small trees. It is common in woodlands, heaths and open forests, especially on infertile sandy or stony soils.

Uses: The first naturalists exploring Australia were astounded by the "topsy turvy" animals they found. It was thought the plant life might prove equally freakish, and native cherry was paraded as proof of this—the little cherry with its seed outside. But Australia's flora is not essentially peculiar, and several foreign as well as Australian plants bear fruits with external seeds—the cashew for example.

Native cherry became a well-known colonial fruit, one often mentioned in accounts of the natural productions of the colonies. Most pioneers seemed to like the slightly astringent taste. Aborigines also ate the fruits.

Australia has another eight *Exocarpos*, all probably edible. They are cypress-like shrubs with the seed outside the tiny sweet fruit. Coast ballart (*E. syrticola*) of southern beaches has a pinkish-white fruit. Leafless cherry (*E. aphyllus*) of southern inland woodlands has a flattened red fruit, and the outback slender cherry (*E. sparteus*) has a red, white, yellow or purplish fruit.

BROAD-LEAVED NATIVE CHERRY

Exocarpos latifolius

OTHER NAMES: Mistletoe tree, broad-leaved ballart

FIELD NOTES: This shrub or small tree has leathery oval leaves, and orange or red fruits 5–8 mm long, attached to a hard external seed. The leaves are leathery, 2.5–10 cm long, with five to seven parallel longitudinal veins. This is the only native cherry (*Exocarpos*) with broad leaves, and the only species found across northern Australia. It grows in most damp habitats, including woodlands, beach dunes, coastal scrubs, and monsoon rainforest along creeks.

USES: Aborigines eat the astringent fruits.

PALE-FRUIT BALLART

Exocarpos strictus

OTHER NAME: Pale ballart

FIELD NOTES: (Not illustrated.) This broom-like shrub, growing 1–2.5 m tall, resembles the native cherry (*E. cupressiformis*), except the fruits, which ripen in summer, are a pale lilac-pink and the foliage is less pendulous. The stems are also more angular, and a darker or brighter green. Pale-fruit ballart grows in open forests and woodlands, including sub-alpine woodlands, often on infertile stony soils in colonies. It is the only *Exocarpos*, apart from *E. cupressiformis*, found close to Sydney and Melbourne.

USES: Aborigines ate the tiny sweet fruits.

M. miquelii

Cycads

families Cycadaceae, Zamiaceae

Other Names: Zamia palms, burrawangs, wild pineapples

Field Notes: Cycads are palm-like plants with leathery fronds arising from a trunk which lies underground or is up to three or more metres tall. The hard leaves are often spine-tipped. The seeds are large (3-8 cm long) and starchy.

Australia has 28 species in four genera.

Cycas (ten species) are found in open forest in northern Australia and eastern Queensland. They are tall cycads with rounded, green or brownish seeds produced on stalks in the crown of the tree.

Burrawangs (*Macrozamia*, 14 species) occur in eastern Australia, apart from *M. macdonnellii* in the centre and *M. riedlei* in the south-west. These cycads have short, often completely underground trunks and red or yellowish seeds produced inside tall, spiny, stalked, pineapple-like cones.

Lepidozamia (two species) are very tall (4-18 m) cycads occurring in wet forests in eastern Australia: *L. hopei* in north Queensland and *L. peroffskyana* in southern Queensland and northern New South Wales. The red or yellowish seeds are produced in tall, spineless, stalkless cones.

Byfield ferns (*Bowenia*, two species) are very different looking, fern-like cycads found in wet tropical Queensland. Aborigines ate the root of *B. spectabilis* but not the seeds of either species, and these cycads are not considered further here.

Cycads grow in colonies in woodlands and forests, mainly on sandy infertile soils. They are primitive plants that hark back 180 million years to the time of the first dinosaurs. Because their leaves poison cattle, they have been exterminated over wide areas.

Uses: Raw or cooked cycad seeds taste palatable but are very poisonous, and during the era of Australian exploration the expeditions of de Vlamingh, Cook, La Perouse, Flinders, Grey, Leichhardt and Stuart were all poisoned, though never fatally. Aborigines knew how to leach out the toxins and the prepared seeds were staple foods. The starchy kernels were cracked or crushed (sometimes cooked first), soaked in water for days or weeks, and the fragments ground to paste, cooked and eaten. Different tribes had different methods of preparation.

In the Northern Territory Aboriginal women also gathered very old seeds from beneath the trees, and by crushing and sniffing were able to identify some as being edible without preparation.

Aborigines could trigger seeding by firing the *Cycas* groves, and the abundant seed so produced sustained enormous social gatherings of hundreds of Aborigines for weeks or even months at a time. Surplus seeds could be ground and fermented in water, providing food months after seeding ended.

Cycad seed fragments have been found at 11 archaeological sites in Australia. At Queensland's Carnarvon Gorge *M. moorei* seeds were found at rock shelters in densities of up to 600 per cubic metre of soil, dating back at least 4,300 years.

In Western Australia the fleshy outer layer of *M. riedlei* was eaten after soaking and aging. On Cape York Peninsula the tubers of young *Cycas* plants were reportedly sliced and pounded and baked into a loaf. Bushmen sometimes extracted cycad starch as flour.

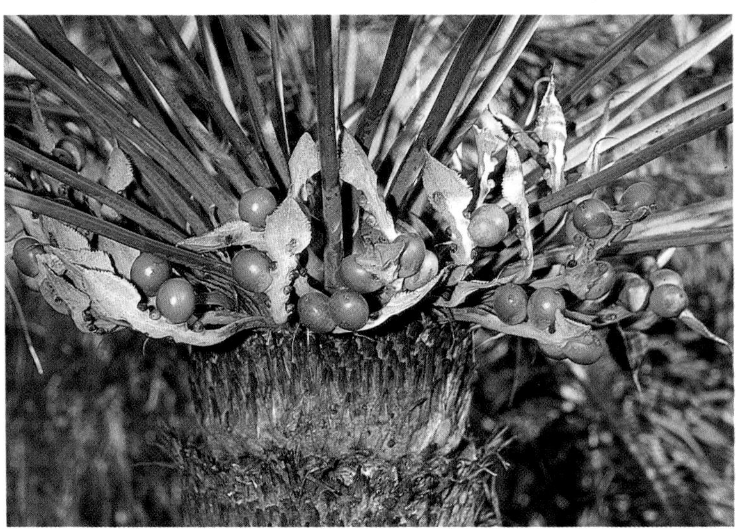

Cycas media

Grasstrees

Xanthorrhoea

Other Name: Blackboy

Field Notes: Grasstrees consist of a woody trunk topped by a crown of tough, wiry, slender leaves. Pale yellow or cream flowers are produced on tall woody spikes. The trunk can be up to 6 m tall, or entirely underground.

Grasstrees are one of the unique features of the Australian landscape. The 28 species grow in sandy or stony infertile soils on ridges and ranges, and in coastal and inland heaths.

Uses: To the Aborigine, the grasstree was an exceptionally useful plant yielding edible starch, nectar, shoots, grubs, gum that served as glue, and flower stalks that could be made into firesticks and spear handles.

The starch was extracted from the upper trunk, as described by James Backhouse in 1843:

> The Aborigines beat off the heads of these singular plants by striking them about the top of the trunk with a large stick; they then strip off the outer leaves and cut away the inner ones, leaving about an inch and a half of the tender white portion, joining the trunk: this portion they eat raw or roasted; and it is far from disagreeable in flavour, having a nutty taste, slightly balsamic.

A different manner of attack was noted by forester G. R. Brown in 1894 at Port Macquarie: "The centre of the leaves at the top is cut deep down, and trimmed of leaves and gum. It is eaten raw, and is very nice and juicy."

South Australian Aborigines harvested the roots, or underground part of the trunk, of *X. quadrangulata,* which Carl Wilhelmi in 1860 described as "by no means tasteless".

Grasstree starch is concentrated food. One Tasmanian sample yielded 41 per cent carbohydrate, more than twice the calorie content of potatoes. The core contains small amounts of protein (3.5 per cent) and traces of vitamin C.

In the nineteenth century, Australian entrepreneurs pioneered techniques for extracting sugar from grasstree cores. An 1876 patent detailed ways of crushing the cores to extract a sweet syrup, which could be rendered into crystal sugar. Fortunately the sugarcane industry prospered instead and the patent was never used.

The harvest of grasstree pith cannot nowadays be condoned, for it kills the tree, and grasstrees grow very slowly. Plucking out the small centre leaves and nibbling the white bases is less destructive. These taste sweet and juicy, or tough and astringent.

The flowers of grasstrees lure honeyeaters and other pollinating birds. Aborigines licked the beads of sweet nectar, or soaked the flowerheads in coolamons to make drinks. Around Bunbury the nectar was fermented into a mildly alcoholic brew.

BANKSIAS

Banksia

OTHER NAMES: Native honeysuckle, candle-sticks

FIELD NOTES: Of Australia's 75 species, 12 were only described and named as recently as 1981. They are all woody shrubs or trees with large, coarse, bushy blossoms, and slender, leathery, often serrated leaves. The flowers of 42 species are yellowish, 11 are orange to brown, the rest are reddish, purplish or mauve-pink. For descriptions of the species, see A. S. George's *The Banksia Book.* The genus commemorates Sir Joseph Banks, the so-called father of Australian botany.

Banksias like infertile soils, and they thrive in heaths and woodlands, especially in southern Western Australia where 58 species occur. Swamp banksia (*B. dentata*) is the only species in northern Western Australia and the Northern Territory, and coast banksia (*B. integrifolia,* illustrated) is the most widespread in eastern Australia, occurring on beaches, mountain forests, granite boulder country and rainforest fringes.

USES: Aboriginal diet was sometimes monotonous, and sugary foods like nectars were greatly appreciated, especially from blossoms as laden as the banksias. These were important foods, especially in Western Australia. Aborigines sucked the sweet beads of nectar, or soaked blossoms in water to make drinks, sometimes allowing natural fermentation to produce an alcoholic brew. "The effect of drinking this 'mead' in quantity was exhilarating," Walter Roth noted in 1903 "producing excessive volubility".

Banksia nectar is a delicious bush food best harvested in the early morning before birds and evaporation deplete the yield. Nectar-laden flowers feel sticky when squeezed.

GOLDEN GREVILLEA

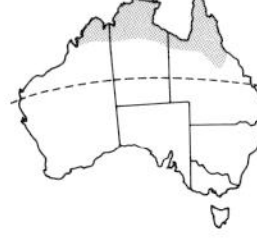

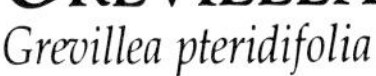

Grevillea pteridifolia

OTHER NAMES: Fern-leaved grevillea, Darwin silky oak, golden parrot tree

FIELD NOTES: Golden grevillea is a tall shrub or small slender tree with showy orange flowers. Leaves are 15-43 cm long and deeply divided like fronds into many slender, tapered segments. The bark is rough and dark. Flowering is from May to October.

Golden grevillea is common in tropical woodlands on low-lying sandy soils. It often grows in colonies along streams and roadsides. It is widely cultivated, and is one of the parents of the popular hybrid Grevillea "Sandra Gordon".

USES: Aborigines sucked copious sweet nectar from the flowers or soaked them in water to make drinks. The flowers are rich in vitamin C, yielding 30 milligrams per 100 grams in one sample tested.

Several other grevilleas in northern, eastern, and central Australia were harvested by Aborigines for their nectar. These include shrubs, trees, even creepers, with very variable foliage and flowers. Any species with flowers that feel wet when squeezed is worth harvesting. In central Australia the yellow and orange flowered honey grevilleas (*G. juncifolia* and *G. eriostachya*) were popular. In eastern Australia the best nectar producer is the silky oak (*G. robusta*), a towering tree of rainforest stream banks in southern Queensland and northern New South Wales with sticky orange flowers. The dwarf silky oak (*G. banksii*), a red or cream-flowered shrub found in eastern Queensland, is another good nectar producer. Both of these plants are very widely cultivated.

Aborigines in northern and central Australia also ate the seeds of some grevilleas.

NATIVE ROSELLA

Hibiscus heterophyllus

OTHER NAMES: Native sorrel, native hibiscus, batham, green kurrajong

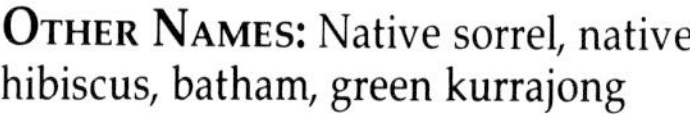

FIELD NOTES: This pretty native hibiscus has showy yellow or white flowers with purple centres. The coarse leaves are dagger-shaped or three-lobed, 5–20 cm long, and the stems are usually prickly. Native rosella is a tall shrub or small tree of shady and swampy eucalypt forests, jungle gullies and rainforest edges.

USES: Rosella jam is made from the acidic buds of the rosella plant (*H. sabdariifolia*), a species of hibiscus, sometimes found growing as a weed in the Northern Territory. Native rosella is closely related, and its sour buds have also been used for jam.

According to A. Thozet, in 1886, Aborigines harvested this plant: "Roots of young plants, young shoots, and leaves eatable." The very sour leaves make a good spinach substitute in Greek dishes and the petals can be eaten in salads. Aborigines also spun the tough stem fibre into string.

There are several other native hibiscuses with edible leaves, buds and roots, found mainly in northern Australia. All are shrubs with showy hibiscus flowers, tough stringy bark, and slightly thickened roots containing mucilaginous starch. As well, the related climbing hibiscus (*Abelmoschus moschatus*) and native rosella (*A. ficulneus*) have edible leaves and roots. No hibiscus is known to be poisonous, and it is probably safe to eat any that taste acceptable. Plants in this family (Malvaceae) may be recognised by their showy petals in fives, and by the way the stamens in the centre of the flower are fused into a central tube.

C. fraseri

KAPOK BUSHES

Cochlospermum

OTHER NAMES: Wild kapok, kapok tree

FIELD NOTES: These are scrawny shrubs or small trees up to 6 m tall with large-lobed leaves and showy yellow flowers. The leaves are 5-15 cm long on long stalks; they are shed during the dry season. The flowers are 5.5-8 cm wide, appearing mainly from May to September; they are followed by egg-shaped capsules, 5.5-9.5 cm long, containing seeds attached to fluffy down that resembles kapok.

There are three species: *C. fraseri* found in Western Australia and the Northern Territory (except eastern Arnhem Land), has leaves with three to seven blunt lobes, resembling maple leaves; *C. gillivraei,* found in eastern Arnhem Land and northern Queensland, has leaves divided with three to seven slender sharp-tipped lobes; and *C. gregorii,* found south of the Gulf of Carpentaria, has its large leaves divided into five separate segments each up to 12 cm long and 3 cm wide.

Kapok bushes grow widely in open woodlands, often on sandstone ridges.

USES: Kapok bush (*C. gregorii*) was a food of Aborigines along the Mitchell River, as Edward Palmer observed in 1884: "The roots of the young trees, which are long and thick, are roasted, when the skin peels off, leaving the edible part white and delicate and well-flavoured." The tasty taproots of young plants of the other species were used in the same way.

Kapok bush flowers (*C. fraseri*) can be eaten as a salad item. They are bland but refreshing in taste.

WHITE BERRY BUSH

Flueggea virosa

OTHER NAMES: White raisin [*Securinega melanthesoides*]

FIELD NOTES: White berry bush is an attractive shrub, growing 1-2.5 m tall, and easily identified by the conspicuous bunches of clean white berries clustered along the stems. The berries are 5-10 mm across, very soft, and contain three or four small seeds. They ripen prolifically between October and April. Leaves are oval, 3.5-8.5 cm long, paler below, with a very finely crinkled surface.

White berry bush grows behind beaches, along streams, in monsoon jungles, and in damp spots in open woodlands. It ranges from northern Australia into New Guinea and across Asia to East Africa.

USES: Aborigines eat the small berries, which have pulp so soft that it "melts" pleasantly in the mouth.

Aborigines also twirled the dried stems to light fires, and some tribes in north-western Australia used the leaves, bark and roots as medicines to treat cuts, sores, rashes, and to soothe the intense pain of catfish stings.

SWEET SANDPAPER FIG

Ficus opposita

OTHER NAME: Sweet fig

FIELD NOTES: The leaves are mainly opposite, oval in outline, with a rough sandpaper feel. The figs ripen from green through yellow to reddish-brown and black, and measure 0.8–2.5 cm wide, becoming very soft when ripe.

Sweet sandpaper fig can be distinguished from the rainforest sandpaper figs (see page 83) by its mainly opposite, not alternate, leaves. Another species, the broad-leaved sandpaper fig (*F. scobina,* also edible) of north-western Australia, differs in having leaves which are broadest in their upper half, and narrower at the base.

This fig is very variable, and some forms look nothing like the example shown here. Around Darwin, and in north Queensland, the leaves are often marked by black tar-like fungal spots. There is no danger of misidentification—any plant with sandpapery leaves has edible fruit.

Sweet sandpaper fig is a sprawling shrub or small tree of stream and rainforest margins, coral atolls and beaches, rocky outcrops and tropical woodlands.

USES: Aborigines harvested the succulent figs, and used the leaves as sandpaper to polish tools. Joseph Banks watched Aborigines using fig leaves in north Queensland while the *Endeavour* was under repair:

> They use shells and corals to scrape the points of their darts, and polish them with the leaves of a kind of wild Fig tree (*Ficus Radulo*) which bites upon wood almost as keenly as our European shave grass usd by the Joiners.

Nowadays, Aborigines in Dampierland use the leaves to scrub pots and to clean spark plugs. In parts of the Northern Territory the inner bark and leaves are infused to treat illnesses.

LADY APPLE

Syzygium suborbiculare

OTHER NAMES: Red bush apple, red apple

FIELD NOTES: This is a very variable tree or shrub with large leaves and big red fruits. The fruits and leaves can look very different from the example illustrated here. Fruits are red or dark pink, often strongly ribbed, 3.5-9 cm wide, with four prominent flaps at the tip, enclosing one big round seed. They ripen in spring and summer and occasionally at other times of the year. The leathery leaves are 7-19 cm long, 4-13 cm wide, glossy or dull green, with blunt or tapered tips. Lady apple is one of the lilly pillies (see page 74).

Lady apple can grow into an impressive, wide-crowned tree, but is often small and scraggly. It is common in lowland open forests, especially on sandy soils and behind beaches.

USES: The fragrantly spongy fruits were a very important food of northern Aborigines. Along coastal dunes in north Queensland, thick groves of the trees have sprung up at old Aboriginal campsites where seeds were dropped long ago. The fruits have a very refreshing taste, and contain up to 17 milligrams of vitamin C per 100 grams. Some tribes use the leaves and fruits as medicines.

The closely related white love apple (*Syzygium eucalyptoides* subsp. *bleeseri*) resembles the lady apple, except its 3.5-4 cm wide fruits are cream or white and lack ribs. It grows in woodlands and along streams in north-western Australia. A related subspecies, *S. eucalyptoides eucalyptoides* found across northern Australia, has 3 cm wide cream, pink or red fruits, and slender leaves like a eucalypt.

Cocky Apple

Planchonia careya

Other Names: Billygoat plum, bush mango, wild quince, mangaloo

Field Notes: Cocky apple is a scrawny tree with fissured bark and large dangling leaves. The leaf blade usually tapers and continues along the length of the leaf stalk, so that there is no sharp division between the leaf and its stalk. The leaf edges are often finely serrated and new leaves are often reddish. The large pink and white flowers have many hair-like stamens. The yellow-green fruit is egg- or pear-shaped, 5–9 cm long, containing two to eight seeds surrounded by stringy flesh, with a ring of flaps and a spike-like style at the tip. Fruits ripen from late spring to autumn.

Cocky apple is a common tree of tropical woodlands and forests, sometimes found growing in sand behind beaches.

Uses: Cocky apple is one of the more important of tropical native fruits, with a stringy flesh that tastes like quince. Aborigines ate them raw or after roasting. The seeds were reportedly eaten as well. On Groote Eylandt fruits are struck to the ground with sticks, and examined to see if the flesh is yellow. White unripe pulp stains and irritates the mouth.

The Department of Defence Support has tested the fruits and found them to be low in thiamine and vitamin C.

Cocky apple is a very important medicine plant in northern Australia. Aborigines infuse the bark and heat or pulp the leaves to treat sores, wounds, stings and other ailments. The bark and roots were also hurled into pools to poison fish, which could then be eaten with impunity.

GREEN PLUM

Buchanania obovata

OTHER NAMES: Wild plum, bush currant

FIELD NOTES: Green plum is a small or medium-sized tree of open woodlands easily recognised by its large, very elongated, mainly oblong leaves. The leaves are dull green, leathery, blunt-tipped, 10-25 cm long, 2-10 cm wide, with very distinct cross-veins. The bark is dark grey and rough. Dangling sprays of tiny creamy-white flowers are followed by yellow-green fruits, 1-1.7 cm long, which consist of a thin layer of flesh surrounding a large seed. They ripen in large numbers from October to January. Green plum is deciduous during the dry season. The new leaves are often reddish.

Green plum is a common tree of tropical woodlands and sandstone country. It often grows alongside billygoat plum and cocky apple.

USES: Aborigines relished the green "plums"which were gathered in large numbers and sometimes dried for future use. The roasted and pounded roots are also reported to be edible.

The green plum was one of the most important medicine plants of Northern Territory Aborigines. Preparations of the inner bark and leaves were applied to sores, sore eyes, and especially to aching teeth.

The related little gooseberry tree (*B. arborescens*), a shiny-leaved tree found in monsoon forests in northern Australia, has purplish edible fruits, which explorer Ludwig Leichhardt made into drinks.

Billygoat Plum

Terminalia latipes subsp. psilocarpa

Other Names: Green plum, vitamin C tree, Kakadu plum, [*T. ferdinandiana*]

Field Notes: This deciduous tree has rough, flaky, grey bark, and large, broad, hairless, blunt-tipped leaves, up to 25 cm long and 20 cm wide, on long stalks. Spikes of tiny cream flowers are followed by pale green or yellow-green, ovoid fruits, 1.5-2.5 cm long, which contain a large woody stone. They ripen from March to June.

Billygoat plum is a common tree of lowland open forest. Its large pale leaves are distinctive.

Uses: The sour fruits, a popular snack food of Aborigines, achieved world fame when a sample tested by the Human Nutrition Unit at Sydney University was found to contain world record amounts of vitamin C (see Chapter 10). The sample scored 3150 milligrams per 100 grams, 60 times the vitamin C content of oranges. There has been talk of cultivating the tree as a vitamin source. Aborigines at Kakadu also eat the pale, almost tasteless gum that oozes from the trunk.

The wild peach (*T. hadleyana* subsp. *carpentariae*) is a very similar-looking tree, distinguished by the covering of very fine hairs on its leaves and fruits, giving them a fuzzy texture. The 1.5-3.5 cm long yellow-green fruits, which ripen from August to October, taste like half-ripe peaches. The gum is also edible. This tree is found growing in open forests in the top end of the Northern Territory, usually on hillsides, plateaus or on stony soils.

A. aneura

ACACIA GUM

Acacia

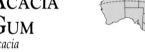

OTHER NAME: Wattle gum

FIELD NOTES: Acacias are shrubs or small trees with fluffy, yellowish flowers and flattened or cylindrical, bean-like pods. The leaves vary greatly between the 800–900 species but are usually either lance- or sickle-shaped with longitudinal veins, or pinnate with numerous tiny leaflets.

USES: Aborigines ate the gums oozing from the trunks and branches of many kinds of wattles. Only the pale gums can be eaten; the darker gums are too astringent.

In Western Australia settler Ethel Hassell told of a gum that oozed in lumps sometimes as big as a door handle. When fresh, she said "it is like pure white sugar-candy, but on exposure becomes a beautiful clear honey colour, and is crisp outside but sweet, soft and sticky inside . . . [Aboriginal women] press the lumps together and make large round balls, about as big as a child's head, and keep it for use".

In Victoria acacia trees were owned by individual Aborigines, who notched them each year to enhance the flow of gum. Gums were often blended with nectars and lerps into drinks.

The gum of Sydney green wattle (*A. decurrens*), a small tree with ridged stems and finely pinnate leaves, was eaten by Sydney Aborigines and settlers. Maiden wrote in 1895:

> It must possess some nutritive value, as instances are on record of the lives of children and others who have been lost in the bush having been sustained by it. Boys sometimes soak it in water to make a thick jelly and sweeten it.

Cherbourg Aborigines even today make jelly by soaking acacia gums in water and sugar and refrigerating.

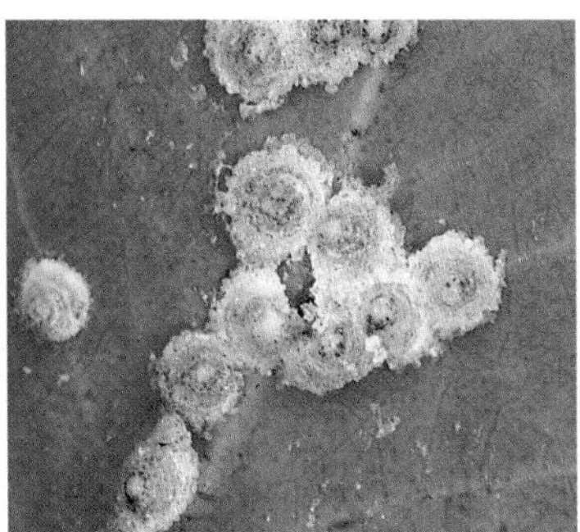

Lerp Glycaspis

Eucalyptolyma *lerps*

Lerp and Manna

Field Notes: Lerps are tiny white protective shields, 2-10 mm long, produced on eucalypt leaves by sap-sucking psyllid bugs. They are made largely of sugars or starches extracted from the sap and excreted by the bug. Lerps resemble tiny limpets, scallops, or tufts of fairy floss.

Other bugs cause trees to "bleed" sugary liquids which evaporate on the leaves or fall to the ground as white flakes called manna (not illustrated). Manna gum (*Eucalyptus viminalis*) and brittle gum (*E. mannifera*) of south-eastern Australia produce copious manna.

Uses: Lerps and mannas were eaten by Aborigines, and the latter were important foods in southern Australia. P. Beveridge claimed in 1884 that:

> an Aboriginal can easily gather 40 or 50 pounds weight of it in a day. As the Aborigines are extremely fond of this sweet substance, during its season they do little else but gather and consume it, and they thrive on it most amazingly.

Another settler, James Dawson, described a manna in western Victoria:

> The natives ascend the trees, and scrape off as much as a bucketful of waxen cells filled with a liquid resembling honey, which they mix with gum dissolved in cold water, and use as a drink.

New South Wales colonists went on genteel outings to collect a manna "sweeter than the sweetest sugar, and softer than Gunter's softest ice-cream". Some mannas are laxative, and were taken as medicines.

Lerp, also called "sugar bread" is harvested by running gum leaves between the teeth. It often occurs alongside a sticky-sweet secretion called "honeydew". The very different mulga lerp is described on page 181.

Bloodwood (E. intermedia)

EUCALYPTS

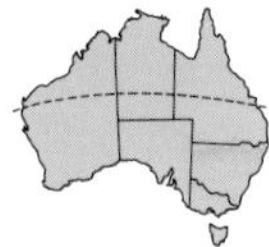

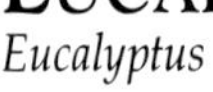

Eucalyptus

OTHER NAMES: (See below)

FIELD NOTES: The eucalypts, popularly known as gum trees, are Australia's most famous trees. The leaves are usually tough, slender, sickle-shaped, drooping and alternate, with a eucalyptus smell. The fluffy flowers are followed by woody gum "nuts". Depending on bark types, the 550 different species are variously called ironbarks, stringybarks, bloodwoods (see page 186), boxes, gums, etc.

USES: Though they dominate Australia's landscape, the eucalypts provided little that Aborigines could eat. The animals harboured by the trees, like lerps, galls, witchetty grubs, beehives, possums and koalas, were usually more important than the trees themselves.

Some outback and northern eucalypts furnish edible seeds, a useful standby food, and others a sweet, starchy root-bark. A couple of West Australian species ooze edible gum, described by settler Ethel Hassell as "a thick, purplish syrup, which is very sweet". Some species, including bloodwoods, secrete so much nectar it can be sucked from the flowers, or soaked in water to make drinks.

Perhaps the most remarkable eucalyptus food is the sugary sap produced by the cider gum (*E. gunnii*) of Tasmania's central plateau. Aborigines drank the sticky sap or stored it in holes until it fermented. According to one old report, "at Christmas time, in 1826, the Lake Arthur blacks indulged in a great eucalyptus cider orgy". Stephen Harris of Tasmania's Department of Parks and Wildlife has suggested cultivating the tree to produce a maple syrup substitute.

L. muelleriferdinandii

PEPPERCRESSES

Lepidium

OTHER NAMES: Inmurta (A), unmata (P)

FIELD NOTES: Peppercresses are small weedy herbs growing 10–20 (rarely 40) cm high. They have slender leaves, often branched or toothed, with a mustard or peppery taste. The seed pods are green, disc-shaped, and have a notched tip. They are spirally arranged around the upper stems, which are crowned by minute flowers.

Mueller's peppercress (*L. muelleri-ferdinandii*) has slender, sometimes branching leaves, mostly less than 3 cm long, and pods 4–6 mm long.

Veined peppercress (*L. phlebo-petalum*) has slender unbranched leaves up to 5 cm long, and pods 6–9 mm long.

Green peppercress (*L. oxytrichum*) has toothed or branched leaves and pods 4.5–5.5 mm long.

Peppercresses sprout in woodlands after winter rains. In dry central Australia they grow only in soils receiving run-off, either in sandy gullies, stream banks, or on gently sloping land well away from watercourses. They are common after heavy rains but do not germinate in dry years.

USES: Peppercresses were eaten by desert Aborigines, who first steamed them in pits between hot stones or layers of reeds, wet grass, or succulent vegetation. The leaves, pods and hammered stems were eaten. Peppercresses were not popular foods but may have been nutritionally important—related cresses are very high in vitamin C.

PIGWEED

Portulaca oleracea

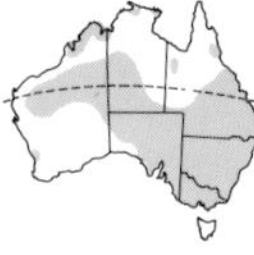

OTHER NAMES: Purslane, munyeroo, wakati (P), lyawa (A)

FIELD NOTES: Pigweed is a succulent, ground-hugging herb with blunt-tipped triangular leaves and small yellow flowers. It grows in clumps up to a metre across but usually much smaller.

Pigweed grows worldwide, mainly as a farm and garden weed, but the large form growing in central Australia is probably a distinct native species. It sprouts on bare soil after rain, often forming thick mats on floodplains. One plant can produce 10,000 seeds.

USES: Tiny black pigweed seeds were a staple food of outback Aborigines. When the stems turned pink the plants were harvested and piled onto hard ground, bark or kangaroo skins. After a few days the seeds fell from the plants and could be gathered up, ground into paste and cooked. Botanist Joseph Maiden mused in 1889: "One would suppose that so small a seed would scarcely repay the labour of collecting", but noted that "the natives get in splendid condition on it." The seeds are a good source of protein and fat.

Aborigines also cooked and ate pigweed roots, and ate the leaves and stems raw, steamed, or ground into paste. Boiled pigweed was the most widely eaten of colonial bush vegetables, and it is sometimes still gathered today. It was eaten by explorers such as Burke and Wills.

Large pigweed (*P. intraterranea*) is a more robust desert plant with petals 12–17 mm long and stems usually more than 1 cm thick. Aborigines ate the thick taproot, which tastes like potato.

PARAKEELYA

Calandrinia balonensis

OTHER NAMES: Ilknguliya (A), Parkilypa (P)

FIELD NOTES: This herb is recognised by its succulent leaves and showy purplish flowers. The five broad petals are 1-1.5 cm long. The leaves are very fleshy and somewhat flattened, mostly 4-10 cm long, with a single sunken mid-vein.

Parakeelya grows on red sandy soils, especially in spinifex country. After heavy rains it may carpet the ground, presenting a massed display of purple flowers. In drier weather it sprouts singly amongst spinifex clumps or in mulga woodland. Parakeelya is easy to identify even when not in flower; there are few other succulent desert plants.

USES: Parakeelya was a standby food of outback Aborigines. The Pitjantjatjarra ate the steamed leaves and roots, and in emergencies ate the raw leaves for their water. The tiny black seeds were sometimes ground to paste and eaten, but were not easy to gather in quantity. They are rich in protein (14.6 per cent) and fat (17 per cent).

East of Birdsville, the leaves were a homestead food early this century. According to Alice Duncan-Kemp:

> Little black children and their white playmates gathered armfuls of Bogil-a-ri or Wild Spinach from the sandy stretches, and red rosy Pigweed for salads, and Mungaroo, a fleshy-leaved plant, or the water-bearing Par-a-keel-ya—all were palatable vegetables when cooked and dressed with seasoning or white sauce.

The closely related *C. remota* with more cylindrical leaves, and the dwarf *C. polyandra*, were also eaten by Aborigines. All *Calandrinia* are probably edible. They are all fleshy-leaved herbs with pink or purplish flowers.

NATIVE MILLET

Panicum decompositum

OTHER NAMES: Native panic, umbrella grass, cooly, tindil, windmill grass, papa grass, kaltu kaltu (P), altjurta (A)

FIELD NOTES: Native millet is a nondescript grass with large, often pale, blue-green leaves, and very spindly, branching seed spikes. The leaves are hairless, up to 50 cm long, and the seed spikes are 30-80 cm, sometimes up to 145 cm tall. The seeds are about 1.5 mm long.

Native millet is a widespread outback grass, found on low-lying ground in most habitats. It is especially common on heavy clay soils, forming dense swards on floodplains, river banks, and in roadside ditches after summer rain. An important pasture grass, it is eagerly grazed by cattle and sheep.

USES: Native millet was a staple food of outback Aborigines, and one of the most important of the native grains. The tiny seeds, produced abundantly in late summer and autumn, are easily husked by hand. They were ground to flour and roasted or baked into nutritious damper. Major Mitchell saw great heaps of this grass pulled by Aborigines for many miles along the Narran River; he rode through a field of millet which ran for nine miles. Native millet is closely related to cultivated millet (*P. miliaceum*).

At least two other species of desert millet were eaten by Aborigines. Hairy panic (*P. effusum*), a very similar-looking grass, has hairy leaves and stems, and grows 20-60 cm tall. Bunch panic (*P. australiense*) is a smaller plant, about 15 cm tall, often with reddish leaves, and with seed heads hidden among the foliage. Its seeds were usually gathered, not from the grass, but from around ant nests.

Woollybutt

Eragrostis eriopoda

Other Names: Wire wanderrie grass, neverfail, wangunu (P), atjira (A)

Field Notes: Woollybutt is a coarse wiry grass with rigid pointed leaves, readily identified at ground level by its white woolly butts, an unmistakable feature. The seed spikelets are 6-22 mm long on stalks 30-60 cm tall.

Woollybutt grows in clumps in large colonies throughout the outback on red sandy soils, often as the dominant grass in mulga woodland. It is very common around the base of Ayers Rock. Plants may live for more than 20 years.

Uses: Woollybutt was one of outback Australia's most important foods, and the most important of the native grains. Its reddish seeds are as small as salt grains, but were highly valued because they are: easy to husk; soft and easy to grind; produced abundantly after rain, staying on the plants for several months; and highly nutritious.

The seed husks were rubbed away by hand, or burnt away with a fire stick, and the seeds were ground between stones, then moistened and roasted in hot sand or ashes. The loaves could be stored for long periods.

There are several other species of *Eragrostis*, all rather similar to woollybutt (however without the woolly butt), that Aborigines harvested for grain. The gathering of other grasses by Aborigines is described on page 15.

TAR VINES

Boerhavia

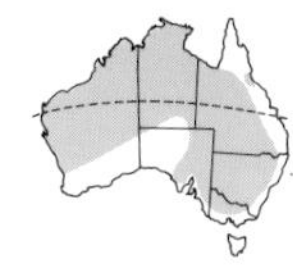

OTHER NAMES: Hog weed, tah vine, spiderling

FIELD NOTES: Tar vines are very common creepers of bare sandy ground. Leaves are triangular or oval, opposite, 1-3 cm long (occasionally to 6 cm), usually with wavy margins and pale undersides. The opposing leaves in each pair are usually unequal in size. Stems are often hairy and sticky and may be up to 3 m long. Flowers are tiny, 1.5-3 mm long, white, pink or purplish, with five joined petals, on long slender stalks. The tiny seed pods are ribbed, and the root is a thickened taproot.

USES: Though inclined to taste bland and fibrous, tar vine taproots were significant foods of Aborigines in central, western and northern Australia. They were usually harvested at the end of the growing season when the taproots were largest. The roots were sometimes eaten raw, but may irritate the throat, especially the bitter skins.

Observations on Aboriginal use of tar vines extends back 150 years, but no one is sure which species were eaten due to confusion over names. Until recently all tar vines were identified as *B. diffusa*, a plant not found in Australia. There are now known to be eight native species, and Aborigines probably ate all the thick-rooted forms. The Kalumburu of Arnhem Land, for example, gave names to four edible kinds—bulga, guranggali, nanggiridji and walinjiri.

In central Australia a colourful hawkmoth caterpillar (*Hyles lineata livornicoides*) found on tar vine foliage was an important Aboriginal food. It has a black tail spike and a greenish and reddish-brown body with black and cream spots and stripes. Roasted, it tastes pleasantly starchy.

ULCARDO MELON

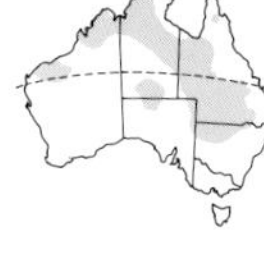

Cucumis melo

OTHER NAMES: Native cucumber, smooth cucumber, native melon, native gooseberry, ilkurta (A)

FIELD NOTES: Ulcardo melon is a small tendril-coiling creeper of flat open ground. Its leaves are 3-7 cm long with a coarse texture, and vary in shape from heart-shaped to deeply lobed. Small yellow flowers are followed by heavy crops of greenish melons, 2-5 cm long, which contain many slender whitish seeds (resembling cucumber seeds) in watery pulp. There are several other kinds of wild melon (see Chapter 7), but these have spiny or very large fleshy fruits.

Ulcardo melon grows after rain on grassy plains, clay soils and flood-prone flats, in Australia, Africa and Asia. It is becoming less common, probably because cattle are eating the fruits.

USES: Surprising though it may seem, the ulcardo melon is the wild ancestor of the rock melon, cantaloupe, and various kinds of Asian melon, all bred from Asian forms of this widespread plant. The ulcardo was a very important food of outback Aborigines, who harvested the watery fruits in great numbers after rain. It was also sampled by explorers Major Mitchell, Augustus C. and Francis T. Gregory, and John McDouall Stuart. Stuart in 1860 attributed the "cucumbers" with relieving his scurvy, and noted: "We boil and eat the cucumbers with a little sugar, and in this way they are very good, and resemble the gooseberry; we have obtained from one plant upwards of two gallons of them."

In Asia, the seeds or seed oil of cultivars are sometimes eaten.

Maloga Bean

Vigna lanceolata

Other Names: Yam, pencil yam, parsnip bean, native bean, ilatjiya (A), kitjutari (P)

Field Notes: Maloga bean is a sprawling creeper with yellow pea flowers. The leaves are 2-8 cm long, carried in groups of three, and vary enormously in shape (see leaf gallery), even among plants growing side by side. The flowers are followed by beans, 2-5 cm long and 4-8 mm wide, smooth or downy but never bristly. Some pods are produced underground like peanuts.

Maloga bean grows mainly in dry watercourses, along creek flats, along roadsides, and at the base of granite hills. It is sometimes a weed of farms on black soils. In northern Australia it grows in woodlands alongside the mung bean (see page 123) with which it can be confused.

Uses: Maloga bean was a staple root crop of inland Aborigines, eaten after the leaves had died away. Aboriginal women dug the very long, slender taproots from sandy river loams and roasted them or ate them raw. The roots are juicy and starchy but bland. Some samples have yielded high levels of magnesium, calcium, iron and thiamine. In northern Australia the delicious green bean pods and underground "toy peanuts" were also eaten raw. On Groote Eylandt these were a food of children.

Aborigines also ate the thin taproot of the related wild cow pea (*V. vexillata*), found in warm coastal forests, and the slender roots of the dune bean (*V. marina*), a very similar-looking, yellow-flowered creeper found on beaches in the northern half of Australia.

BUSH POTATOES

Ipomoea

OTHER NAMES: See below

FIELD NOTES: These are vigorous vines or twining shrubs with dark green, slender, oval or heart-shaped leaves and showy trumpet flowers with purplish centres.

Weir vine (*I. polpha*) is a large ground-trailing creeper with rosy pink or blue flowers and large leaves, 5-18 cm long, found in north Queensland, southern Queensland (St George, Roma) and the central Northern Territory (at Tea Tree).

Goolaburra (*I. calobra*) is a vine that twines up mulga and other acacias in south-western Queensland and Western Australia. It has pink flowers with a darker centre and leaves up to 12 cm long.

The desert yam (*I. costata*) begins life as a vine, but eventually develops into a woody-stemmed shrub with wiry, vine-like stems. It has pink flowers with red or purple centres, broad leathery leaves, 4-9 cm long, and grows on spinifex sand plains in the Northern Territory and Western Australia.

USES: Bush potatoes are closely related to the sweet potato, and they also produce big tasty tubers, sometimes as large as a human head.

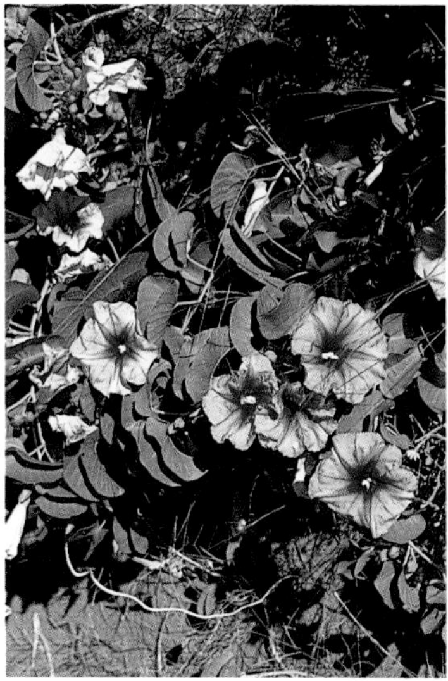

I. costata

These were highly favoured staple Aboriginal foods. They lie deeply buried up to 2.5 m from the base of the plant. Women sometimes detected the tubers by thumping the ground with sticks and listening for a hollow sound. Tubers of the desert yam and weir vine were sometimes traded between tribes. It has been suggested that the weir vine's rarity and patchy distribution reflect overharvesting in the past.

In the woodlands of northern Australia there are several related species of *Ipomoea* with showy trumpet flowers and edible tubers once eaten by Aborigines.

M. viridiflora

Bush Bananas

Marsdenia australis, M. viridiflora

Other Names: Silky pear, native pear, native potato, doubah [*Leichhardtia*]

Field Notes: These wiry twining vines have slender opposite leaves, 4-12 cm long, 4-5 mm wide, which ooze milky sap when plucked. Cream or greenish flower clusters are followed by greenish drooping pods, 4-10 cm long, which split to release fluffy wind-borne seeds. The foliage is usually sparser than shown here.

The two species look alike. Doubah (*M. australis*) occurs in south-western, southern and central Australia, and native potato (*M. viridiflora*, previously *M. leptophylla*) in the east and north. They can be confused with gargaloo or monkey vine (*Parsonsia eucalyptophylla*), a larger vine with a thick woody stem, dense foliage, and larger (to 30 cm), drooping, usually yellowish, leaves.

Bush bananas twine over shrubs, fences, and sometimes over the ground in inland woodlands. Native potato also grows in brigalow scrubs.

Uses: The young roasted pods make an exquisite vegetable, tasting like the finest baby squashes or zucchini. Doubah pods were very popular Aboriginal foods, as noted by explorers Sturt and Mitchell. Exceptionally rich in thiamine, they yield up to 2900 micrograms per 100 grams—more than ten times that of any cultivated food.

Native potato produces strings of watery tubers, sometimes as big as a human head; these were also eaten by Aborigines or sucked to allay thirst. Doubah has only very slight tubers, but these were eaten by desert Aborigines, along with the leaves, flowers, shoots and seeds of this very versatile food plant.

NATIVE PEAR

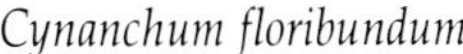

Cynanchum floribundum

OTHER NAMES: Desert cynanchum, ngiltha

FIELD NOTES: Native pear is a shrub (rarely a bushy creeper or climber), with opposite, pointed leaves, and green wiry stems that ooze milky sap when broken. The leaves are 3-9 cm long, and vary from broad and heart-shaped to very slender (as illustrated). The white or greenish star-shaped flowers are followed by tapered green pods, 3-5 cm long, which split when ripe to release brown flattened seeds attached to silky hairs.

Native pear grows in red sand dune deserts in the south, usually in deep sand on the crest of dunes, but in the north grows in or near watercourses. It usually sprouts as a broad shrub about 1 m high consisting of many vertical, vine-like stems arising from ground level. Although very variable, it cannot be confused with any other plant.

USES: Desert Aborigines ate the raw unripe pods, which have a pleasant vegetable flavour. Older pods and leaves were steamed and eaten. The inner stem bark was spun into sturdy string.

The related climbing purple star or purple pentatrope (*Rhyncharrhena linearis*, previously *Pentatropis kempeana*) is a slender, wiry, outback vine with milky sap, starry greenish-yellow to purple five-petal flowers, very slender leaves, and very slender, pointed, bean-like pods, 8-22 cm long, 5-8 mm wide, which split when ripe to release fluffy air-borne seeds. Aborigines ate the pods, buds, flowers, leaves and roots. The pods are a good source of vitamin C.

WILD TOMATOES

Solanum

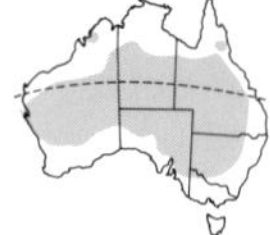

OTHER NAMES: Potato bushes, tomato bushes (and see below)

FIELD NOTES: Wild tomatoes are shrubby, often prickly plants growing 15-100 cm tall, with pale green or greyish, finely furry leaves, which feel velvety. Flowers are purple or lavender with five joined petals and yellow centres. The fruits closely resemble small unripe tomatoes. When ripe they are pale yellow, often with a green, purple or brown tint, 1-2 cm wide, and contain many pale seeds. The bush tomato (*S. chippendalei*) and kakarrta (*S. diversiflorum*) differ in having 2-3 cm fruits and black seeds. The dried fruits often resemble raisins.

Some species of wild tomato have bitter poisonous fruit, and sometimes the edible kinds produce bitter inedible fruit. The best guide to edibility is taste.

Wild tomatoes sprout in colonies in most outback habitats, especially after rain or fire. They are common on roadsides and floodplains.

USES: Wild tomatoes were staple foods of outback Aborigines. Desert botanist Peter Latz has described one species, called desert raisin (*S. centrale*), as "probably the most important of all the central Australian plant foods". Its fruits were eaten fresh (though overeating them causes illness) or dried. Dried fruits were sometimes ground with water into paste and moulded into kilogram cakes that were stored.

RUBY SALTBUSH

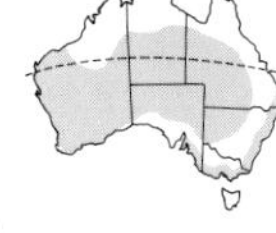

Enchylaena tomentosa

OTHER NAMES: Barrier saltbush, Sturt's saltbush, plum puddings, berry cottonbush, creeping saltbush

FIELD NOTES: Leaves are grey-green and sausage-shaped, 0.6–2 cm long. The stems are brittle and often hairy. The fruits are tiny, shiny buttons about 5 mm across, usually bright red in colour, sometimes orange or yellow, and ripen at almost any time of year.

Ruby saltbush is one of outback Australia's most common shrubs, occurring in almost every habitat. It is a soft, squat, tangled, ground-hugging shrub, often found growing at the base of trees or fence posts. It can grow up to 1.5 m tall, but some forms are much smaller, only about 20 cm tall.

There is a small form that grows along seashores on salt marshes and rocky headlands. It can be confused with wallaby bush (*Threlkeldia diffusa*), a southern seashore shrub which differs in having leaves that taper to a sharp point and having purplish-red, barrel-shaped fruits. Both leaves and fruits of wallaby bush are edible.

USES: Tiny, salty-sweet fruits of ruby saltbush were once a snack food of desert Aborigines. In the MacDonnell Ranges the fruits were soaked in water and the liquid drunk like sweetened tea. Sometimes the dried fruits were soaked and eaten. The fruits are tiny, consisting mainly of the large seed, and their food value is very slight. Nowadays they are mainly eaten by children although honeyeaters and pigs also eat them.

Ruby saltbush is one of many fleshy-leaved plants that can be boiled and eaten as a vegetable; it was harvested by explorer Charles Sturt in the 1840s.

CURRANT BUSH

Carissa ovata

OTHER NAMES: Turkey bush, baroom bush, blackberry

FIELD NOTES: This scrawny shrub has broad opposite leaves, mostly 1-2 cm long, and pairs of straight spines along stems that ooze milky sap when cut. The fruits are oval, brownish-black, 1-1.5 cm long, and ripen any time of year.

Currant bush grows in arid woodlands, dry rainforests, seashore scrubs, and along rainforest margins and mangrove fringes.

USES: Aborigines ate the berries, which taste like juicy dates, though they are full of gritty seeds. In 1945 a stockman lost for days in bush near Clermont said the berries saved him from hunger and thirst.

KONKERBERRY

Carissa lanceolata

OTHER NAMES: Conkle berry, conker berry

FIELD NOTES: (Not illustrated.) Perhaps only a variety of currant bush, konkerberry has smaller reddish-black berries, 0.5-1.5 cm long, and much narrower, very slender, dull greyish-green leaves (see leaf gallery), 2-4 cm long. It otherwise fits the description of currant bush. It is a prickly, sprawling or upright shrub of dry woodlands, growing up to 3 m tall.

USES: Aborigines eat the tasty berries, which contain traces of vitamin C. Dried berries, gathered from beneath the bush, were sometimes soaked and eaten.

NITRE BUSH

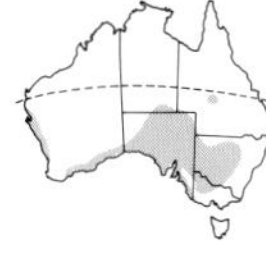

Nitraria billardieri

OTHER NAMES: Dillon bush, wild grape, karumbil, [*N. schoberi*]

FIELD NOTES: Nitre bush forms a large, sprawling, sometimes spiny shrub, 0.5-3 m tall. The fruits are red, purple or yellow, 1-2 cm long, contain a large, pointed seed, and ripen in summer and autumn. The slender fleshy leaves are 1-4 cm long. Plants in south-western Australia grow upright, have broader leaves, and may represent a separate, undescribed species.

Nitre bush grows on beaches, dunes and coastal headlands, and inland on saltbush plains, claypans, river flats, overgrazed pastures and degraded farmlands. It sometimes forms impenetrable thickets.

USES: The fruiting of nitre bushes was a time of feasting for Aborigines in southern Australia. The botanist Carl Wilhelmi in 1860 described bushes:

> so full of fruit, that the natives lie down on their backs under them, strip off the fruit with both hands, and do not rise until the whole bush has been cleared of its load.

Emus are also very fond of the fruits, and flocks of up to 80 have been seen feasting at the bushes. Their droppings may contain more than a thousand of the seeds, which are designed for dispersal by emus. The bird's stomach strips a layer from the seeds that inhibits germination.

Nitre fruits taste like salty grapes. Some crops are more palatable than others, and explorer J. M. Stuart suggested their cultivation, an idea championed recently by CSIRO researcher J. C. Noble. He had samples of fruit canned, but found they tasted bitter and astringent and yielded little vitamin C.

DESERT LIME

Eremocitrus glauca

OTHER NAMES: Limebush, native kumquat, desert lemon

FIELD NOTES: Desert lime is a spiny shrub or small slender tree with slender leaves, 0.3-1 cm wide, and small citrus fruits. The white flowers, produced in spring, are followed in late spring and summer by yellow or green fruits, 1-2 cm long, which resemble tiny lemons, having a porous rind and sour juicy centre.

Desert lime grows in inland woodlands and brigalow scrubs, especially on heavy clay soils, where it may form prickly suckering thickets on land cleared for grazing. In South Australia it is only found near Carrieton, north-west of Port Augusta.

USES: Desert lime is a true citrus, and its sour, tangy fruits make an excellent substitute for lemons in drinks and conserves. Country people gather them for marmalade, and Aborigines ate them raw. Eaten whole like kumquats, they are very refreshing, but may leave a bitter aftertaste. Explorer Ludwig Leichhardt made them into a dish "very like gooseberry-fool" and observed that "they had a very pleasant acid taste".

Desert lime has been used experimentally in horticulture: it crosses easily with cultivated citrus trees, yielding viable hybrids, and its drought resistant root-stock is suitable for grafting. Colonial botanist Baron von Mueller noted the likelihood of its improving in culture, and to its fitness for being grown in arid land.

WESTERN BOOBIALLA

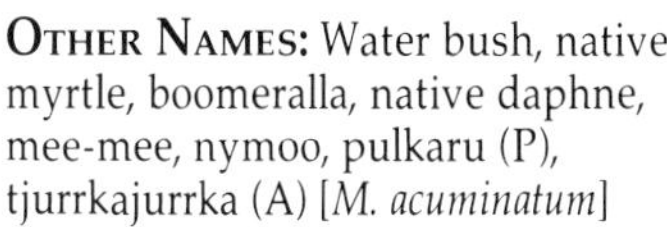

Myoporum montanum

OTHER NAMES: Water bush, native myrtle, boomeralla, native daphne, mee-mee, nymoo, pulkaru (P), tjurrkajurrka (A) [*M. acuminatum*]

FIELD NOTES: This nondescript shrub, growing 1-4 m high, has slender leaves, 3-10 cm long and 5-10 mm wide, which taper to a point. The small white flowers have five blunt petals. Identifying features are the purplish spots inside the flowers, and the bright purple or dark pink fruits, which measure 6-8 mm across and contain one large stone.

Western boobialla is a very widespread outback shrub, found in most habitats, but especially near water, and in pastures heavily disturbed by grazing, where it is a serious woody weed. It also grows along seashores, where it can be confused with the coastal boobiallas (see page 38).

USES: Aboriginal children snack upon the small fruits, which are unpalatable unless completely ripe. Most of the fruit consists of the large stone. A white gum oozing from the stems was used as glue, and the leaves were a medicine of the Arrante.

The related ellangowan poison bush (*M. deserti*), known in some areas as turkey bush or dogwood, is a similar outback shrub with yellow fruits about 6 mm long, white unspotted flowers, and smaller leaves, 2-5 cm long, 3-6 mm wide, which are poisonous and have a bitter, burning taste. Major Mitchell wrote that this bush "put forth sweet and edible fruit".

NATIVE CAPERS

Capparis

OTHER NAMES: Native pomegranates, wild oranges, bumble trees, splitjacks

Pickled capers are the prepared buds of a Mediterranean shrub, *Capparis spinosa*. This plant produces pungent mustard oils as its defence against predators, and these give capers their spicy flavour. Identical oils occur in plants of the mustard family, to which capers are related.

The "Mediterranean" caper ranges widely in Africa and Asia, and early botanists were puzzled to find it growing in northern Australia, and thought it to be an introduction of early man. We now know it to be a native variety (var. *nummularia*) adapted to deserts. Buds of this variety make excellent pickled capers, but have rarely been used. The shrub is an important Aboriginal food, however, furnishing sweet fruits rich in thiamine.

Australia has another 17 or so native capers, all bearing fruits, and all apparently edible. They were important Aboriginal foods. One species, *C. canescens*, has mustard oils in its rind, and was used in Queensland last century as a mustard substitute.

Native capers are shrubs or trees of very variable appearance, but their flowers and fruits provide easy identification. The flowers are usually large and showy, but flimsy and fragile, with four broad, white petals and numerous, long, hair-like stamens. The fruits have many seeds surrounded by sticky pulp, like a passionfruit or pomegranate. They hang singly on long, often curved, prominently jointed stalks. They can be reddish, orange, yellow, green or black when ripe, are sometimes covered in bumps and lumps of gum, and may be attacked by grubs and ants. White caper butterflies often flutter above the foliage, and their caterpillars and pupae rest upon the leaves. Another distinctive feature is the hooked spines often found along the stems, especially of young plants. Most capers undergo a metamorphosis during growth, beginning as a wiry scrambler with tiny leaves and maturing into a shrub or tree with larger leathery leaves.

Apart from the capers featured here, there are species found in eastern rainforests, along northern beaches, and in tropical and subtropical woodlands. Some are rare and little known. All five outback species are described here.

NARROW-LEAF BUMBLE-TREE

Capparis loranthifolia

FIELD NOTES: (Not illustrated.) This is a shrub or small tree closely resembling wild orange but with narrower leaves. These are 0.8–1.8 cm wide, with five to eight pairs of cross-veins, compared with three to four pairs in wild orange.

USES: Aborigines eat the big reddish-brown fruits.

Wild Orange

Capparis mitchellii

Other Names: Bumble tree, bimbil, mondo, native pomegranate, mpiltjarta (A), mpultjati (P)

Field Notes: This is a small compact tree with dull green, rigid, leathery, blunt-tipped leaves, 2-6.5 cm long and 1.2-3.8 cm wide. Showy cream or white flowers are followed by rounded fruits, 3-7.5 cm wide, yellowish-green and fragrant when ripe, often warty or oozing gum, and which are carried on long, curved, jointed stalks. Wild orange grows scattered in woodlands throughout the outback. Its dense crown is conspicuous.

Uses: Aborigines relished the fruits, which taste like passionfruit with a kerosene aftertaste.

Caper Bush

Capparis spinosa var. *nummularia*

Other Names: Wild passionfruit, flinders rose, split arse, splitjack, nipan

Field Notes: This is a scrawny, spiny shrub about 1-1.5 m tall with soft, dark green, oval or rounded leaves, 1.5-4 cm long. The green fruits are egg-shaped and strongly ribbed, about 3-4 cm long. When ripe they turn yellow, apricot or reddish and split open. Fruits ripen in summer and autumn.

Caper bush grows on river flats, hillsides, woodlands, paddocks, rocky beaches and Barrier Reef islands. Its dark foliage is distinctive.

Uses: Aborigines eat the passionfruit-flavoured pulp.

NIPAN

Capparis lasiantha

OTHER NAMES: Splitjack, nepine, maypan, honeysuckle, alurra (A)

FIELD NOTES: Young plants are spiny creepers with zigzag stems. Mature plants are squat, tangled shrubs, sometimes found creeping around tree trunks or fence posts. Leaves are dull green, leathery, blunt-tipped, 2.5-8 cm long, on spiny zig-zag stems. Fruits are egg-shaped, 2.5-5 cm long, and turn yellow or pale green and split when ripe, in late summer and autumn. Nipan is common in woodlands, rocky hillsides, paddocks and roadsides.

USES: The sticky fruit pulp is pleasant eating.

NORTHERN WILD ORANGE

Capparis umbonata

FIELD NOTES: (Not illustrated.) This slender tree is distinguished from other capers by its drooping branches and long, slender, drooping leaves. Leaves are dull green, 10-23 cm long, 0.7-3 cm wide, sometimes sickle-shaped. Fruits are globular, 3-4.5 cm wide, with a thick yellow or reddish rind, on a very long, curved, jointed stalk. Fruits ripen in the warmer months.

Wild orange grows in open woodlands and forests, sometimes on rocky slopes and often along river flats.

USES: The fruits are highly regarded by Aborigines, who sometimes ripen the green fruits in hot sand.

DESERT FIG

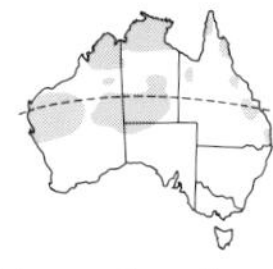

Ficus platypoda

OTHER NAMES: Small-leaved rock fig, small-leaved Moreton Bay fig, ili (P), yili (P), pilka (P), tjurrka (A)

FIELD NOTES: This is a very large sprawling shrub or small multi-stemmed tree of rocky hills and ranges. The leaves are dark green, smooth and leathery, 6-10 cm long and 1-4 cm wide. The stems ooze milky sap when broken. The rounded figs are 1-1.6 mm wide, consist of a thin layer of sweet flesh surrounding the gritty seeds, and turn from yellow to dark red when ripe. They ripen any time of year following rain.

In the outback the desert fig grows only upon rocky hills, gorges and ranges, beyond reach of fire. It can be seen around the base of Ayers Rock, in the Olgas, and around the Devil's Marbles. One Queensland variety, known as the small-leaved rock fig, grows into a huge tree of coastal rainforests. It is found largely on rocky outcrops, even sprouting on the convict ruins of St Helena Island in Moreton Bay. Other varieties grow in rocky places in northern Australia.

Figs as a group are described in the Rainforest section on page 80.

USES: The desert figs were, and still are, an important and popular food of desert Aborigines. They are rich in potassium and calcium. Dried fruits were sometimes gathered from beneath the trees and ground into an edible paste, an important drought food. The shrubs grow very conveniently beside most permanent waterholes in the desert. Settlers stewed the fruits into jam.

The desert fig was a sacred totemic tree of some desert tribes; anyone harming the plants could be killed.

SANDALWOOD

Santalum lanceolatum

OTHER NAMES: Plum bush, bush plum, wild plum, cherry bush, blue bush, plumwood, kupata (P)

FIELD NOTES: Sandalwood is a scrawny shrub or very small, slender tree with lazy drooping branches and characteristic blue-grey foliage. The leaves are opposite, varied in shape, 2-9 cm long, with blunt or pointed tips. The fruits are 7-14 mm wide, disc-shaped at the tip, contain one large rounded stone, and turn red then purplish-black when ripe.

Sandalwood is a widespread shrub found scattered through woodlands, brigalow scrubs, woodlands behind beaches and rocky hillsides. In drier country it is restricted to watercourses. It is common around the base of Ayers Rock and along gully lines in the Olgas.

USES: Sandalwood fruits are reasonably sweet and palatable, and were a popular food of Aborigines. They have been found to contain between 2 and 16.4 milligrams of vitamin C per 100 grams. Dried fruits found beneath the plants were sometimes reconstituted with water. The desert-dwelling Warlpiri were also reputed to have roasted and ground the seeds to make an edible paste.

Sandalwood, and closely related trees of the same name, have aromatic wood, much in demand in China for incense. Sandalwood timber was once exported from Queensland through Rockhampton and Thursday Island. The aromatic oil was also blended with other sandalwood oils as a medicine. Some Aboriginal tribes used the leaves or inner wood as medicines, and the fruit juice as a dye.

DESERT QUANDONG

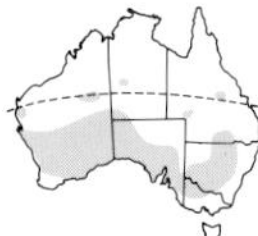

Santalum acuminatum

OTHER NAMES: Native peach, sandalwood, katunga, burn-burn, mangata (P)

FIELD NOTES: This shrub or small tree has paired olive green leaves, 3-9 cm long and 3-15 mm wide, on drooping stems. The shiny, bright red (sometimes yellow) fruits are 2-3 cm wide and contain a pale knobbly stone. They ripen in spring.

Quandongs grow in woodlands on sandy and stony soils, usually as scattered shrubs or trees. They are becoming scarce in the Northern Territory, apparently from camel grazing.

USES: Without doubt, the quandong is outback Australia's favourite wild fruit. Aborigines relished the scarlet acidic pulp, which is high in vitamin C. It sometimes served as a staple food for short periods. Dried fruits found under the trees were reconstituted with water, or fashioned into cakes which were stored.

Settlers gathered the fruits for pies, jellies and jams, even drying the excess, like the Aborigines, for future use. Explorer E. J. Eyre declared that the fruit "makes excellent puddings or preserves, for which purpose it is now extensively used by Europeans". The fruit is nowadays served in wild food restaurants.

Quandongs also have oily edible kernels which Aborigines sometimes ate. Some trees produce sweet, almond-flavoured kernels, others have distasteful kernels.

CSIRO has proposed the cultivation of the quandong as a desert fruit crop. It is very tolerant of soil salinity. But unfortunately the seeds are very difficult to germinate, and the plant is a root parasite, requiring a host tree from which to draw nutrients.

Emu Apple

Owenia acidula

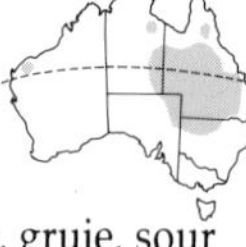

Other Names: Colane, gruie, sour plum, sour apple, gooya, moalie apple, native peach, wild peach, mooley apple

Field Notes: The emu apple resembles the quandong, except its bright green leaves are shiny and pinnate, like those of pepper tree (*Schinus*), and the fruits are a duller, purplish-red with paler speckles. They are 2–4 cm wide and contain a very large stone. Broken twigs ooze a milky sap and the leaflets are 2–5 cm long.

Emu apple grows into an attractive, small or medium-sized tree of outback woodlands, often found in clumps arising from the suckers of a single tree.

Uses: Explorer Major Mitchell was the first European to taste emu apples, which he found had "an agreeable acid flavour". He observed that "it was evidently eaten by the natives as great numbers of the stones lay about".

The dark pulp is rich and tangy, though very acidic, resembling a very sour plum in flavour. They ripen only after falling to the ground. Settlers used them to make jam. One colonial traveller in Queensland described the fruit as "grateful to the taste of a thirsty traveller in these hot arid regions".

Some writers have complained that emu apples are highly distasteful. They have perhaps sampled fruits straight from the tree, though it is also possible that in some areas, such as the Northern Territory, the fruits are inedible. Bushmen in Queensland's Channel Country swear that the local trees never fruit.

Bitter Quandong

Santalum murrayanum

Other Name: Ming

Field Notes: (Not illustrated.) Bitter quandong closely resembles the desert quandong but the reddish fruits are very bitter, and the grey or silvery leaves are very small, 1.5–4 cm long and 2–4 mm wide.

Bitter quandong grows in mallee and other woodlands.

Uses: Aborigines reportedly ate the seeds and roasted bark of the roots. According to one writer the root and bark were made into a stupefying drink.

A. dictyophleba

ACACIA SEED
Acacia

OTHER NAME: Wattles (and see below)

FIELD NOTES: The outback acacias are shrubs or small trees with fluffy yellowish flowers. Leaves are usually slender, sometimes sickle-shaped or spine-tipped. Seeds are produced in a row in pods which are either cylindrical or flattened, and which, while green, often resemble bean pods.

Acacias occur throughout Australia in all but the very wettest habitats. More than a hundred species are found in the outback, where they are the dominant shrubby vegetation. Mulga (see next page) is the most important edible species. Acacia gums are depicted on page 152.

USES: Acacia seeds were very important foods of outback Aborigines, eaten green or ripe, and the seeds of more than 30 species were eaten. Green seeds were eaten like peas after roasting the green pods in a fire. (The raw green seeds and pods are intensely bitter.) The very hard, ripe seeds were either ground, moistened and roasted as damper, or were roasted first then ground between stones into a paste tasting like peanut paste. Ripe seeds were gathered from pods on the trees, from the ground below, or from the nest entrances of seed-gathering ants. After good rainfalls the seed was so abundant that large social gatherings of Aborigines could be supported. Acacia seeds are extremely nutritious, yielding protein levels of 18–25 per cent, and sometimes high levels of fat. The most important outback acacias are mulga, dogwood (*A. coriacea*), *A. cowleana*, *A. dictyophleba*, strap wattle (*A. holosericea*), witchetty bush (*A. kempeana*), and bramble wattle (*A. victoriae*).

Groves of mulga trees (Acacia aneura)

MULGA

Acacia aneura

OTHER NAMES: Wanari (P), kurrku (P), ititja (A), artitja

FIELD NOTES: Although mulga is the most common tree in outback Australia, it is not always easy to identify. Some forms (illustrated) have cylindrical leaves only 1 or 2 mm wide; others have broad flat leaves up to 1 cm wide with longitudinal veins. The fluffy cylindrical yellow flowers are followed by oblong flattened pods, 1.5-7cm long, 0.4-1.5 cm wide, containing very hard, shiny black seeds about 5 mm long. Mulga grows as a large shrub or small tree. Helpful identifying features are the greyish colour of the foliage; the upright, not drooping, crowded branches; and the tendency of mulga to grow in dense pure stands on red soils.

USES: To the Aborigines, mulga was one of central Australia's most useful plants, supplying abundant edible seed, gum, honeydew, "apples", and a very strong timber used for tools.

Seeds were gathered from the trees or swept up from the ground below, roasted, and ground to a paste said to taste like peanut paste. Desert botanist Peter Latz estimates a mulga yield of at least a hundred kilograms of seed per hectare. By this reckoning the mulga in the Northern Territory could feed a quarter of a million people.

Mulga stems occasionally ooze glistening lumps of gum when attacked by insects (see illustration page 152). This gum, candy-hard on the outside, syrupy sweet within, was eagerly eaten.

Mulga is often attacked by a sap-sucking bug called mulga lerp (*Austrochardia acaciae*) which produces shiny red lumps along the stems and oozes a sweet liquid called honeydew. Aborigines broke off the stems and sucked up the honey by drawing the stems between the lips, eventually resulting in cracked and bleeding lips. The lerp itself is not edible. Mulga lerp is also found on stems of witchetty bush (*A. kempeana*), a similar-looking shrub with spherical fluffy flowerheads, better known for the delicious witchetty grubs found in its roots.

Mulga "apples" are sweet edible galls produced by a small wasp larva which lives inside. The apples are marble-sized, and covered in small scattered lumps, features that distinguish them from other kinds of (inedible) mulga galls. The galls taste like dried apple, and were a popular treat of Aborigines and bushmen. Colonial botanist Joseph Maiden described them as "a great dainty".

At least three kinds of mistletoe parasitise the branches of mulga, and Aborigines ate the fruits of these: mulga mistletoe (*Lysiana murrayi*) has very slender green leaves and pink to dark red fruits; grey mistletoe (*Amyema quandang*) has broad greyish leaves and fruits; and pale-leaf mistletoe (*Amyema maidenii*) has broad grey leaves and yellow berries.

Mulga apple

Mulga lerp (Austrochardia acaciae)

The famous honeypot ant (*Melophorus inflatus*) usually builds its nest beneath mulga. Aboriginal women dug deep into the sand to find the honey-swollen ants, which were esteemed as a great treat.

Mulga wood was the most important outback timber for tools such as boomerangs, spear blades, woomeras, digging sticks, and shields. The timber can be recognised by its pale outer wood and very dark heartwood.

WEEPING PITTOSPORUM

Pittosporum phylliraeoides

OTHER NAMES: Berrigan, bitter bush, butter bush, cattle bush, cheesewood, cumby cumby, locketbush, meemei, native apricot, native willow, poison-berry tree, snotty gobbles, western pittosporum

FIELD NOTES: This is a shrub or small tree with pendulous branches and leaves, and distinctive orange or yellow capsules, 1-2 cm, which split open when ripe to expose the sticky, reddish, bitter-tasting seeds. The leaves are slender, bright green, 3-10 cm long.

Weeping pittosporum is common and widespread in many habitats. In very dry regions it is restricted to watercourses.

USES: Aboriginal use of plants sometimes varied widely from place to place. Weeping pittosporum was to some tribes a source of food and to others a medicine, while still other groups, like the Pintubi and Arrante, considered it useless—a mere shade tree. Even as a food its use varied from place to place—some Aboriginal groups ate the gum that oozed from wounded branches, and others reportedly ground the seeds to flour. The seeds taste intensely bitter, and it is not surprising that most tribes disregarded them.

Weeping pittosporum had a widespread reputation for healing, though the parts of the plants used, like the ailments treated, varied from place to place. Seeds, leaves or wood were used to treat colds, cramps, sprains, eczema and itching.

Kurrajong

Brachychiton populneus

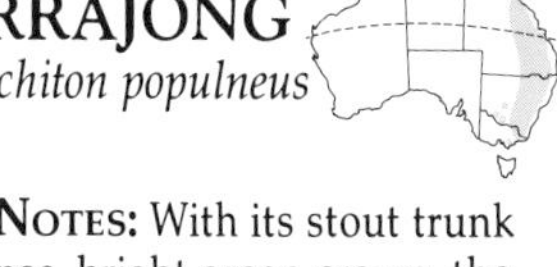

Field Notes: With its stout trunk and dense, bright green crown, the kurrajong is a conspicuous outback tree. The shiny, drooping leaves are tear-shaped or three- or five-lobed, with long tapering tips that flutter in the breeze. Bell-shaped, red-streaked flowers are followed by leathery, boat-shaped pods about 5-7 cm long.

The kurrajong is a common tree of sandy inland plains, extending to the eastern slopes of the Great Dividing Range. It grows in woodlands, open forests, and dry rainforests. It is often left in cleared paddocks as fodder for cattle.

Uses: Kurrajong seeds are remarkably nutritious, comprising about 18 per cent protein and 25 per cent fat, and yielding high levels of zinc and magnesium. Popular Aboriginal tucker, they were eaten raw or roasted after removal of the irritating yellow hairs. They taste pleasantly nutty when roasted over a high heat. The swollen roots of young trees were also eaten.

Several other *Brachychiton* have edible roots (when young) and seeds. They have pods like the kurrajong, but dissimilar leaves. The northern kurrajong (*B. diversifolius*), distinguished by its big, smooth, oval- or heart-shaped leaves, and the red-flowered kurrajong (*B. paradoxum*), a small tree identified by its big hairy leaves and hairy pods, are found in northern woodlands. The desert kurrajong (*B. gregorii*) of western deserts has very slender three- or five-lobed leaves. Also, the bottle tree (*B. rupestris*) of inland Queensland has edible roots when young, and the widely cultivated rainforest flame tree (*B. acerifolium*) has nutty edible seeds.

CORKWOODS
Hakea

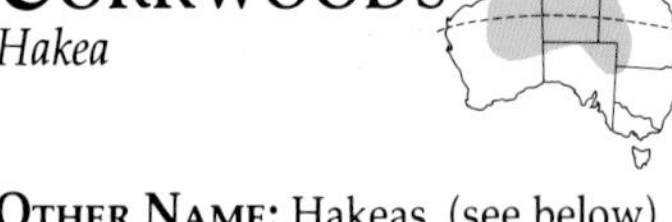

OTHER NAME: Hakeas, (see below)

FIELD NOTES: These are scrawny trees with thick, very deeply fissured bark and showy, creamy-yellow flowers like those of a grevillea. The tough leaves are very slender, either flattened and leathery, or cylindrical like pine needles, or divided into many wiry spiny segments. The woody seed pods are 2-4 cm long, with a bird-like beak at one end.

Corkwoods are common woodland trees of outback Australia. At least four species were harvested for their nectar: the northern corkwood or bootlace oak (*H. chordophylla*), the fork-leaved corkwood or straggly corkbark (*H. eyreana*), the long leaf corkwood (*H. suberea*); and *H. macrocarpa*. Genus *Hakea* includes about 130 species, found throughout most of Australia, and known generally as "hakeas". Aborigines probably harvested the flowers and seeds of other species.

USES: Corkwood flowers secrete copious amounts of tangy nectar, and this was highly valued by desert Aborigines. The flowers were sucked, or steeped in water in large numbers to make a sweet drink. The flowers of the northern corkwood produced a blackish drink that was said to be slightly alcoholic. Aborigines also ate the seeds of the fork-leaved corkwood.

Corkwoods were important Aboriginal medicine plants. The burnt bark was applied to burns and open sores, or was mixed with animal fat to make a healing ointment.

In the deserts of southern Australia Aborigines were able to survive in waterless regions by draining watery sap from the roots of the related silver needlewood (*H. leucoptera*).

WILD BAUHINIA TREE

Lysiphyllum gilvum

OTHER NAMES: Bean tree, bohemia, [*Bauhinia gilvum*]

FIELD NOTES: This is a large outback tree with characteristic cleft leaves (see leaf gallery), resembling butterflies in flight or the leaves of cultivated bauhinia trees. The leaves are 2-4 cm long with several longitudinal veins radiating from the base. The flowers have five hairy, cream or reddish petals, with long protruding stamens, opening in spring, followed by flattened woody bean pods. The bark is dark grey and rough.

The wild bauhinia is found mainly along outback watercourses, where it is often the dominant tree.

USES: Alice Duncan-Kemp, who settled at Mooraberrie Homestead west of Winton, recorded the Aboriginal use of this tree:

> When rains draw near, the old gins about the homestead wander up and down the creeks scoring or cutting the bark with a tomahawk. After rain they visit the scored trees and gather the *minni*, a thick sweet sap which exudes from the cuts like jellied honey-coloured gum; a great delicacy among the blacks, who eat it straight from the tree.
>
> White children consider *minni* an excellent substitute for sweets, when cooked in a baking-dish with a little sugar and water for a couple of hours, and sprinked with roasted nuts or *Mungaroo* from the roots of the nut-grass [*Cyperus bulbosus*].

Alice also told of Aborigines making semi-intoxicating drinks by mixing pounded bauhinia flowers with native honey or honeypot ants, the liquid being left to ferment for eight or ten days. In the Cloncurry area the nectar was sucked from the flowers or soaked in water to make drinks.

WESTERN BLOODWOOD

Eucalyptus terminalis

OTHER NAMES: Pale bloodwood, kulcha, arrkirnka (A)

FIELD NOTES: This is a small to medium-sized tree with pale, rough, scaly or fissured bark that extends up to the twigs. The leaves are dull, pale green, alternate, on stalks 0.8–1.7 cm long. The flowers are cream, the stems often carry large galls, and the gum nuts are urn-shaped and relatively large and thick, 1.5–4 cm long.

Western bloodwood occurs as scattered trees in open woodland, on sand plains and dunes, on river flats, and on stony ridges. With its pale, fissured bark and pale yellowish-green foliage, this is an easily recognised tree.

USES: A coccid bug (*Apiomorpha pomiformis*) parasitises the stems of this tree, producing large woody galls which were a popular Aboriginal food, known nowadays as "bush coconuts" or "desert apples". These are rough globular structures the size of golf balls or apples. Old galls are dark brown, brittle and inedible; fresh galls are very pale, and can be broken open to expose a soft white grub. Both the grub and the white flesh lining the gall are sweet and edible, and the latter both looks and tastes much like the flesh inside green coconuts.

The western bloodwood is a source of other foods. Aborigines sucked white lerps (see page 153) from the leaves, and soaked the flowers in water to make a sweet drink. Hives of bee honey ("sugar-bag") are often found inside the trunk, and water can be drained from the roots. A dark red gum-like kino found oozing from the trunk was a very important remedy for sores. It was mixed in a little water and applied to sores, wounds and burns.

7

Introduced Food Plants

Native plant foods are the subject of this book, but so many exotic food plants now run wild in Australia that it is regrettably necessary to draw attention to some of these. All of the following plants can be found growing in disturbed bushland where they are likely to be confused with native plants. As well, there are hundreds of edible weeds found on farms, roadsides and wastelands; these are depicted (with recipes) in my earlier book, *Wild Herbs of Australia & New Zealand*. When harvesting exotic plants, take care not to spread about the seeds.

Stinking passionfruit (*Passiflora foetida*) is a soft vine with three-lobed or heart-shaped velvety leaves, and thin-skinned, yellow-orange fruits about 2 cm wide, enclosed in "spidery" bracts. It grows in lowland open forests and on beaches from the Richmond River, New South Wales, north to the Hamersley Range, Western Australia. A popular

The yellow skinned fruit of the stinking passionfruit (Passiflora foetida) *is very tasty.*

Aboriginal food, it is widely thought to be native, but in fact comes from Latin America. The cultivated passionfruit (*P. edulis*) is also a widespread weed in eastern Australia.

The pie melon or bastard melon (*Citrullus lanatus*) is the wild form of the watermelon and cultivated pie melon. It is an African creeper with lobed leaves, yellow flowers, and greenish melons up to 15 cm long which, when ripe, have dark brown seeds with black stripes. (The similar but poisonous colocynth [*C. colocynthis*] has yellow seeds and intensely bitter pulp.) Widespread in outback Australia, the pie melon was an important food of colonists. It is still made into jam by country people.

European blackberries (*Rubus fruticosus, R. discolor, R. ulmifolius, R. laciniatus*) are very common along streams, gullies and roadsides in south-eastern and south-western Australia. They are large scrambling shrubs with prickly arching stems, serrated leaves grouped mainly in threes and fives, and juicy black berries of good flavour. It is believed that colonial botanist Baron von Mueller spread the seeds in the forests of Victoria, where they are now the state's worst weed.

The tiny purplish-black fruits of lantana (*Lantana camara*) and bitou bush or bonseed (*Chrysanthemoides monilifera*) are edible. These sprawling shrubs of coastal forests are invidious noxious weeds.

Prickly pears (*Opuntia*) are common and well-known cacti with broad

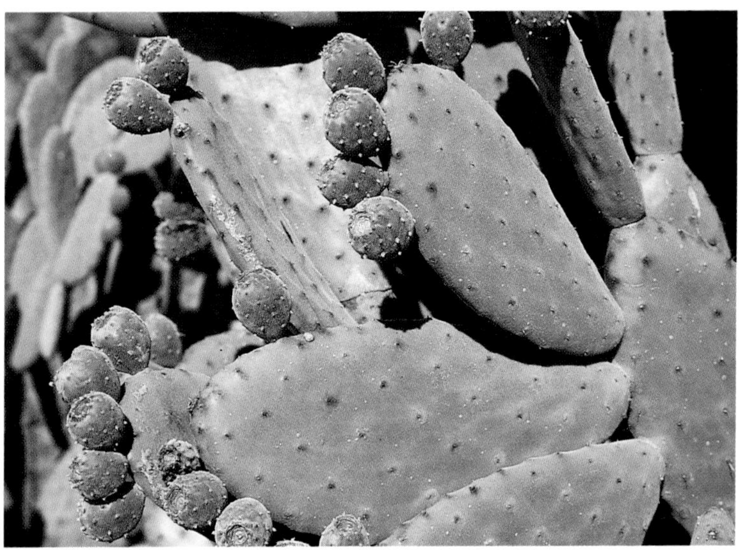

The drooping tree pear (Opuntia vulgaris) *has sticky edible fruits.*

flat pads covered in tufts of irritating hairs, yellow or orange-red flowers, and red or purple juicy fruits. The sticky fruits have a slimy sweet pulp, though great care should be taken to first remove the tufts of irritating hairs on the skin. Prickly pears are found in woodlands and on beaches in eastern and southern Australia. There are 20 or so species in Australia, all native to Latin America.

African boxthorn (Lycium ferocissimum) *berries ripen in the warmer months.*

African boxthorn (*Lycium ferocissimum*) is a spiny shrub with thick, bright green, blunt-tipped leaves, white or lilac flowers, and orange or red berries 5–12 mm long resembling tiny tomatoes. It grows on beaches and black soil plains in eastern and southern Australia. I have eaten the slightly bitter berries without ill-effect. The native boxthorn (*L. australe*), a species found in southern inland Australia, has smaller, fleshier leaves but otherwise looks similar: Aborigines ate the fruits.

The bush lemon (*Citrus limon*) is the wild form of the lemon, distinguished by its thick green knobbly skin. It grows along streams near rainforests in southern Queensland and New South Wales. The juice is sour but flavoursome. Other common foods gone wild include the olive (*Olea europaea*) in South Australia and the guava (*Psidium guajava*) in Queensland.

Wild tobacco *(Solanum mauritianum)* is a large shrub or small tree of eastern rainforest clearings with big, soft, velvety, grey-green leaves, violet starry flowers, and clusters of dull yellow globular berries. The berries were eaten by colonists, but are usually too bitter to eat.

The chinee apple or Indian jujube (*Zizyphus mauritiana*) is a tall shrub or small tree with thorny zigzag stems, oval leaves with pale undersides, and yellowish or red fruits, about 2.5 cm wide, resembling small apples. Chinee apple was introduced to northern Australia by Chinese immigrants, and now grows wild around Townsville, Darwin, and many other northern towns. It is the wild form of the Chinese date.

The African tamarind (*Tamarindus indica*) was introduced to the Northern Territory hundreds of years ago by Indonesian *bêche-de-mer* fishermen, and now grows scattered along isolated beaches. It is a tree with small pinnate leaves, and long dangling pods consisting of a brittle shell enclosing a dark brown, very sour, sticky pulp. Aborigines ate the pulp, which tastes like dried apricot.

Coconuts (*Cocos nucifera*) have been planted on many beaches in northern Australia, and were cultivated by islanders in Torres Strait. The milk and flesh of the green, ripe and sprouted nuts are edible, as is the inner heart at the top of the trunk, though harvesting it kills the tree.

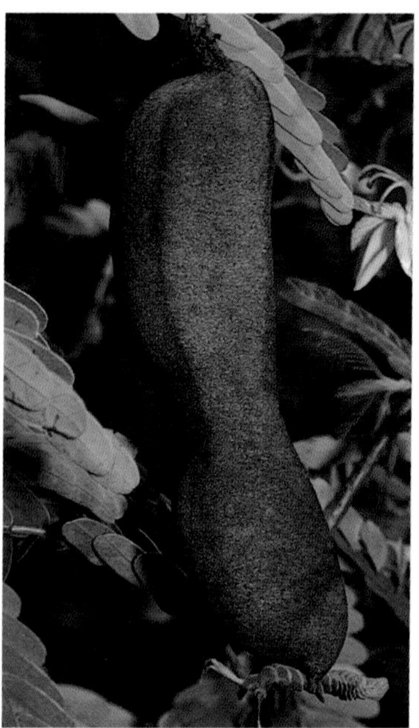

The dangling pod of the African tamarind (Tamarindus indica) *is brown or black and brittle when ripe.*

Wild tobacco (Solanum mauritianum) *berries are sometimes too bitter to eat.*

8

MUSHROOMS

Australia's mushrooms are poorly known, and it is impossible to provide anything like a complete guide to the edible kinds. This brief chapter describes only a few of the more common or easily identified species.

Most of the mushrooms that sprout in Australian lawns and paddocks are well known European or American species. Their spores were probably brought in accidentally in potting soil long ago, or they may have arrived naturally, either blown in by winds or on the feathers of migrating birds. Their eating properties are well known, and they can usually be identified from English and American books.

The fungi found in forests and other native habitats are more problematic. Many have not been scientifically named, and very little is known of their use by Aborigines.

The familiar field mushroom (*Agaricus campestris*) found on lawns is an introduced species closely related to the mushroom of commerce. It has lustrous whitish caps, and gills which darken with age from

The forest mushroom (Agaricus langei) *when fresh makes excellent eating.*

pink to chocolate brown. The similar but larger horse mushroom (*A. arvensis*) has a double ring on the stalk and gills which are first cream, not pink. The native forest mushroom (*Agaricus langei*) is also very similar, but its gills are paler pink when young, they are separated from the stalk by a gutter, and the ring around the stalk is a distinctive papery skirt. it grows on lawns and in forests.

When collecting these mushrooms, avoid similar fungi with any of the following features: yellow, white or greenish gills; caps with a yellowish centre and a tendency to stain blue or green when crushed; or caps that are thimble-shaped when young and smell of disinfectant when stored in a plastic bag.

The only deadly mushroom known in Australia, the infamous death cap (*Amanita phalloides*), has a yellowish or greenish cap and white gills. An introduced species, it grows in parkland in south-eastern Australia beneath exotic trees. It is especially dangerous because, like related poisonous *Amanita* it has no unpleasant warning taste. Most *Amanita* have white gills and a prominent stem ring.

Puffballs (*Lycoperdon*) are another mushroom group found sprouting on lawns after rain. Looking at first like blobs of putty, they soon turn dry and brown and release spores through a hole at the top. They are edible while still firm and white. Some species found in fields are enormous—up to a third of a metre across. Any puffball with consistently white flesh is safe to eat.

Puffballs (Lycoperdon) *are common lawn fungi.*

Of the few fungi that Aborigines were recorded eating, the most famous is the so-called "blackfellow's bread" (*Polyporus mylittae*). The bumpy whitish cap has wavy edges and sprouts from a thick underground stalk. The underside of the cap is white and porous. It grows near dead trees in rainforests and wet and dry eucalypt forests, is widespread, but rarely seen. The underground stalk is attached to an edible tuber, weighing up to 20 kilograms, with a texture like compacted boiled rice. Aborigines rated this a delicacy.

In Tasmania (and probably Victoria) Aborigines ate the beech fungus (*Cyttaria gunnii*). The related *C. septentrionalis*, found in southern Queensland and New South Wales, is also edible. These insipid, usually orange fungi resemble golf balls, and sprout in clusters in spring on the trunks of antarctic beech trees (*Nothofagus*).

Also eaten by Aborigines was the desert truffle (*Elderia arenivaga*), a rare fungus located by finding tiny cracks in the soil.

Bracket fungi are the plate-like or ear-like fungi that sprout on trunks, stumps and logs. Most are woody and inedible, but the following species are soft and edible.

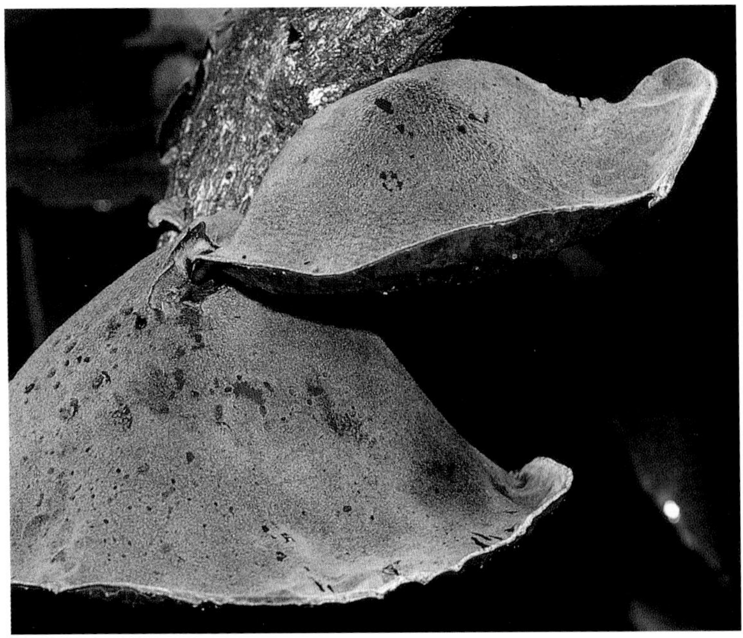

Hairy jew's ear (Auricularia polytricha) *is a rubbery, wafer-thin bracket fungus with a pleasant taste.*

The beefsteak fungus (*Fistulina hepatica*) looks and feels like a slice of red meat. The upper side is reddish and moist, the underside is composed of white tubular pores.

The hairy jew's ear (*Auricularia polytricha*) is a thin rubbery bracket fungus with a velvety grey upper surface and a purplish-brown underside. It sprouts on decaying rainforest wood after rain, and becomes tough and horny when dry (but can be reconstituted). The delicate flavour suits Chinese cooking.

Other bracket fungi with rubbery or sloppy textures are worth trying. Any that taste acceptable are unlikely to be poisonous.

Jelly fungi (*Tremella*) are another group found on wood in damp forests. They are usually yellow, orange, white or brown, with a sloppy gelatinous structure and no obvious shape, though some kinds resemble tiny antlers or stalactites. Not much is known about them but they appear to be harmless, having a pleasant jelly taste.

The stalactite fungus (*Hericium clathroides*), another species sprouting on rotting wood in wet shady forests, consists of hundreds of hanging white stalactites. It has a delicate flavour.

Coral fungi (*Ramaria*) are pink or orange, fleshy, multi-stalked ground fungi, resembling corals or cauliflowers. Not all species are considered edible.

There are many other edible species, including ink-caps (*Coprinus*), and boletes (*Boletus*).

An edible yellow jelly fungus (Tremella) *growing on a rainforest tree after rain.*

9

Foraging and Cooking

The wild food forager should always exercise care and consideration. Fruits are the foods of birds and mammals so some of each crop should be left untouched. Soil upturned by digging for tubers can create habitat for introduced weeds. Aborigines were sometimes wasteful, but we cannot afford to be—there is too little wilderness left.

Tuberous plants can rarely be tugged intact from the ground. Instead, a narrow hole should be dug to the base of the roots with a small spade or a steel bar. Shallow, stem-attached tubers, like those of small sedges and orchids, can often be uprooted by driving a slender hand spade or steel tent peg down beside the plant and tilting upward. Lilies and orchids have small fragile tubers easily lost to careless digging. Tubers often run deeper than expected, especially in sandy soils.

Toxic tubers and seeds can be leached by grating them and suspending the gratings for several days in a regularly-used toilet cistern. This technique can be dangerous and is not recommended.

Arda (Cartonema parviflorum) *has an edible tuber that can be easily dug with a small spade.*

Wild leaves usually taste better when taken from young healthy plants that have not flowered or seeded. Very salty or bitter leaves may need to be boiled in two changes of water to remove the taint.

It is not always easy to tell when fruits are ripe. Some species only ripen after they fall to the ground. Ripe fruits are usually soft, easy to peel and pluck, and often odourous. Unripe fruits are hard and distasteful, and often contain tannins that dry the mouth and make the teeth feel fuzzy. Some fruits have distasteful skins that should be removed.

RECIPES

Australia never developed a traditional cuisine based upon native ingredients. To compare with the United States' turkeys, blueberries and corn, we have only the macadamia nut, a crop first developed in America. Apart from a few country recipes for quandong pie and native jams, native Australian cuisine remains a novelty.

The rediscovery of an Australian wild foods' cuisine began in earnest in 1984, with the opening in Sydney of Rowntrees, Australia's first wild foods' restaurant. Following the establishment of wild food suppliers, other restaurants have added native food recipes to their menu. The most popular ingredients are bunya nuts, macadamias, wattle seeds, New Zealand spinach, and native fruits, especially desert quandong, brown pine, Davidson's plum, and cherry alder, which are made into sauces. The fruits and nuts are mostly gathered from suburban trees. Restauranteurs and suppliers have in some cases contrived to create (or retrieve from obscurity) colourful names for the foods—New Zealand spinach is marketed as "Warrigal greens" and brown pine as "Illawarra plum".

Cooking with wild foods requires no special skill, preparation or prescription. Excellent jams can be made using the traditional jam method outlined below. Fleshy seashore leaves are ideal for pickling, and a simple pickle recipe is supplied here. For recipes employing leaves the reader is referred to my earlier book on wild herbs.

Jams and Jellies

Some native fruits are exceptionally sour and need to be stewed or made into jam to be properly appreciated. The Davidson's plum, rainforest lime, and Burdekin plum in particular make excellent tangy jams. The standard jam recipe gives good results and is given here.

> Remove the seeds and boil the chopped fruits in as little water as possible until soft, then add an equal volume of sugar. Boil for up to an hour or so until the setting point is reached. To test

for setting, place a drop of the jam onto a chilled saucer and wait a few minutes. If the surface wrinkles when nudged with a finger the jam is ready. Pour into sterilised jars and allow to cool.

Very sweet fruits, like those of figs and raspberries, may need less sugar—say 3/4 cup per cup of pulp. Some fruits, the roseleaf raspberry, for example, are low in pectin and need lemon juice to promote setting—1 lemon to 4 cups of pulp is suggested. Slightly unripe fruits contain the most pectin and these are best for jams.

Some fruits, like native currants and bush lawyer, do not separate easily from their seeds and are better made into jelly.

Boil the fruits in very little water for up to an hour until soft, then strain through muslin, overnight if possible. Add sugar to the pulp, cup for cup, then cook as for jam.

A very tangy jelly can be made from the tiny acidic fruits of the leafless currant bush (Leptomeria aphylla).

Spanikopita

The following recipe gives excellent results using New Zealand spinach, native rosella leaves, sea purslane, and many garden weeds.

Fry a diced onion in butter then add 2–3 (tightly packed) cupfuls of wild leaves and fry lightly. Add the mix to 200 grams of crumbled fetta cheese and 2–3 eggs, add pepper and chopped black olives (optional), and use as a filling under layers of buttered filo pastry. Bake in a moderate to hot oven for about half an hour until the pastry browns.

Pickles

Sydney's earliest colonists made pickles from the fleshy leaves of seablite and the stems of samphire. Caper buds, pigweed leaves and stems, and other fleshy seashore plants are also suitable.

> First boil the plants briefly in a large volume of water to remove some of the salt. Do not let the leaves become limp and soggy. Capers and pigweed can be used raw.
>
> The pickling solution should be prepared from fresh whole spices and good cider or malt vinegar. To 4 cups of vinegar add 6 cloves, 6 whole peppercorns, 2 tablespoons of allspice, a cinnamon stick, a tablespoon of mustard seeds and a garlic clove.
>
> Bring the vinegar and spices slowly to the boil, remove from the heat and leave covered for 2 hours. Pour the vinegar into the jars containing the leaves or capers then seal and store in darkness for several weeks before using.

Seaweeds

Australia's seaweeds are closely related to those found in Japan, but their culinary potential remains largely unexplored. The following culinary suggestions are based on overseas recipes.

As a rule, green and brown seaweeds are good for savouries, and red seaweeds for desserts. Seaweeds must always be scrubbed clean and soaked in fresh water before use. Strips of brown and green seaweeds can be added to stir-fry Chinese dishes or baked with eggs in pies. Green and brown seaweeds can be added to salads and pickles. Sea lettuce (*Ulva*), a common seaweed of rock pools resembling a small soggy lettuce, is one that can be used in these ways. Some of the red seaweeds can be made into jelly.

> Wash and sun-dry the seaweeds, boil them in a little water, then strain through a cheesecloth. The liquid sets to a jelly constituency in the fridge. To flavour the jelly reboil it with prepared coconut milk and sugar, and return to the fridge. Alternatively, soak the dried seaweed in water, tie it in cheesecloth and place it in a pan of milk simmering and sugar, which should set if refrigerated.

Species of *Euchema, Gracillaria* and *Hypnea* can be used in this way. Most are red or whitish.

10

WILD FOOD NUTRIENTS

Until recently little was known of the nutrients in Australian wild foods. Large scale testing was begun in earnest only in the 1980s, by teams from the University of Sydney and the Federal Department of Defence Support.

The University team, headed by Dr Jennie Brand,* tested many Aboriginal foods, mainly from northern and central Australia, including ants, locusts and other animal foods. The Defence Support team, headed by Mr Keith James** and supplied by Major Les Hiddens (the "Bush Tucker Man"), tested more than 200 northern plants—those likely to be encountered by Australian soldiers repelling an Indonesian invasion. James tested extensively for ascorbic acid (vitamin C) and thiamine, as the "human body does not store these vitamins to a great extent and they are rapidly depleted when dietary sources become deficient".

The analyses show an enormous range of nutrients. Many foods gained very low scores, and some very high scores, of all the nutrients tested.

Some plants proved exceptionally rich in thiamine (vitamin B1) and ascorbic acid. Candlenut (4700 micrograms per 100 grams in one sample) and doubah (2935 micrograms per 100 grams) yield more thiamine than any commercial food, the richest of which is wheat germ (2200 micrograms).

One sample of billygoat plum tested for vitamin C by Brand's team scored 3150 milligrams per 100 grams, apparently a world record, and one that generated worldwide publicity. Other samples yielded rather lower (2850 milligrams and 406 milligrams), but still exceptional, vitamin C levels. The cultivation of this tree as a vitamin source is now under consideration.

Other rich vitamin C sources include the hairy yam (223 milligrams as recorded by Brand, though James recorded none) and the fruits

* *Human Nutrition Unit of the Department of Biochemistry and the Commonwealth Institute of Health, University of Sydney.*

** *Food Science Section, Armed Forces Food Science Establishment, Department of Defence Support, Scottsdale, Tasmania.*

of *Cynanchum pedunculatum* (119.2 milligrams) and great morinda (89.8 milligrams in one sample).

James concluded: "Perhaps the most striking result is the high levels of several essential trace metals and minerals found in many indigenous plant foods." Very few western fruits or vegetables produce anywhere near 25 per cent of the recommended daily allowance (RDA) in a 100 gram sample, James noted, yet of 35 wild foods, "14 were found to provide at least 25 per cent of RDA per 100 grams, of potassium". Seventeen contained 25 per cent of the RDA of iron, 10 of calcium and seven of zinc. Exceptionally mineral-rich foods included the nuts and pith of the boab tree (*Adansonia gregorii*) of Western Australia, and the seeds of red kurrajong, sea almond, peanut tree, and some grasses.

Table 1 presents nutrient analyses of some common foods, reproduced by courtesy of Keith James (except for kurrajong and edible spike rush, taken from Brand's work). Table 2 shows typical western foods for comparison.

The figures in Table 1 should be viewed with caution for they represent analyses of only single plant samples, not average figures, as in Table 2. Enormous variations have appeared among different samples. Of three candelnut samples tested by James, two recorded extremely high thiamine levels, in excess of 4000 micrograms per 100 grams, but a third registered only 0.043, an insignificant figure. Two great morinda fruits scored 89.9 and 22.6 milligrams of vitmain C per 100 grams respectively, and variation in the billygoat plum and hairy yam has already been noted.

The variation is probably due to a combination of factors—deficiencies in testing techniques (small sample sizes, deterioration of samples, varying ripeness of fruits), and natural variations among the plants attributable to genetics and environment. James, in commenting on the variation among great morinda samples, noted that they were all collected from the same area at the same time. The nutritional figures quoted throughout this book are mostly taken from Brand and James' work, and are subject to the same qualifications.

TABLE 1 COMPOSITION OF AUSTRALIAN WILD FOODS

Plant	Water %	Protein %	Fat %	Energy kJ/100g	Thiamine (vit. B1) mcg/100g	Ascorbic Acid (vit. C) mg/100g
Candlenut (*Aleurites moluccana*)	24.4	7.8	49.9	2426	4236	—
Native ginger (*Alpinia coerulea*)	89.3	0.69	0.14	ND	Tr	Tr
Kurrajong (*Brachychiton populneus*)	5.6	18.1	24.7	1455	ND	ND
Caper bush (*Capparis spinosa*)	64.2	7.2	7.4	869	692	2.0
Konkerberry (*Carissa lanceolata*)	70.3	1.9	2.1	601	44	2.0
Davidson's plum (*Davidsonia pruriens*)	91.5	0.4	1.7	130	—	—
Hairy yam (*Dioscorea bulbifera*)	71.8	3.57	0.07	453	—	—
Long yam (*Dioscorea transversa*)	73.5	1.73	ND	ND	74	1.6
Long yam (*Dioscorea transversa*)	85.9	1.34	0.13	235	—	—
Blue quandong (*Elaeocarpus grandis*)	69.7	1.1	1.1	ND	—	1.6
Tall spike rush (*Eleocharis dulcis*)	45.4	2.6	0.8	648	ND	ND
Cluster fig (*Ficus racemosa*)	87.9	0.8	0.34	276	—	—
Doubah (*Marsdenia australis*)	71.7	4.1	0.7	506	2935	4.0
Cabbage palm (*Livistona australis*)	86.4	2.52	0.36	ND	53	Tr
Great morinda (*Morinda citrifolia*)	88.3	1.04	ND	ND	58	89.8
Great morinda (*Morinda citrifolia*)	80.0	1.17	0.57	ND	10	22.6
Cocky apple (*Planchonia careya*)	78.9	2.8	0.7	402	—	2.5
Burdekin plum (*Pleiogynium timorense*)	75.9	1.0	2.6	376	—	5.0
Brown pine (*Podocarpus elatus*)	61.0	0.9	0.29	728	—	11.0
Pigweed (*Portulaca oleracea*)	85.5	5.9	0.2	232	131	—
Bracken (*Pteridium esculentum*)	66.5	0.96	0.3	ND	24	—
Roseleaf raspberry (*Rubus rosifolius*)	85.4	1.37	0.31	ND	32	Tr
Leichhardt tree (*Nauclea orientalis*)	78.5	0.96	1.0	ND	Tr	2.5
Leichhardt tree (*Nauclea orientalis*)	73.4	ND	ND	ND	—	29.0
Wild tomato (*Solanum chippendalei*)	78.7	1.14	0.61	430	243	—

ND not determined • Tr Traces • —indicates the vitamin content was below measurable levels

Unlike Table 2 which shows averaged figures, these are the compositions of individual samples.
Considerable variation is likely in the wild—see for example the figures for great morinda and Leichhardt tree fruits.
The doubah figures refer to the pods, the pigweed samples refer to the whole plant, and the bracken samples refer to the rhizomes.
(Reproduced by permission of K. W. James, Officer in Charge, Food Science Section, Armed Forces Food Science Establishment)

TABLE 2 COMPOSITIONS OF AUSTRALIAN HOUSEHOLD FOODS

Food	Water %	Protein %	Fat %	Energy kJ/100g	Thiamine (vit. B1) mcg/100g	Ascorbic Acid (vit. C) mg/100g
Apple	84.9	0.3	0.3	221	36	6
Carrot	89.2	0.9	0.2	150	61	6
Grape	81	0.7	0.4	276	54	4
Lettuce	94.5	1.3	0.3	71	55	13
Onion	90.2	1.2	0.2	146	30	10
Orange	86.1	0.9	0.2	188	84	50
Peanuts	5.4	26.5	47.9	2372	930	Tr
Pears	84.1	0.4	0.3	234	22	4
Potatoes	77.8	2.0	0.1	334	110	14
Tomato	93.7	1.0	0.3	87	60	22

Source: Commonwealth Department of Health. Kilocalories have been converted to kilojoules as follows: 1 kcal = 4.184 kJ. Note that these figures are based on raw foods; some thiamine and ascorbic acid are lost during cooking.

Recommended dietary allowances for adults in Australia are 30 mg per day of ascorbic acid and 800–1100 micrograms per day of thiamine.

11

Bush Survival

Despite Australia's vast size, it is difficult to become seriously lost in the bush. Most watercourses, coastlines and roads soon lead to civilisation.

In the rare event of major mishap, the need to find food is usually secondary. It is more important to have water, shelter from extremes, and to be positioned where help will arrive. People can survive many weeks without food, although under cold or other extremes the body expends its reserves much faster.

It is usually easy to find a few wild food snacks, and these can be invaluable for lifting morale, but to sustain oneself by foraging is very difficult. Aborigines won an easy living from the land, but only because they knew its resources intimately.

The challenge in bush survival is to satisfy hunger because the problem posed by wild foods is finding enough of them: the quest is for calories. Vitamins and minerals are not essential to short-term survival. Protein and fats are more important, but are easily obtained from insects and lizards.

The highest energy foods are animals, seeds, tubers, and inner shoots. It is impossible to live upon leaves, seaweeds or mushrooms, and only a few fruits, such as desert quandong and muntari, can be relied upon as staples.

Seeds are the most energy-rich plant foods but they are not practical survival foods. Smaller seeds need to be efficiently harvested and processed whilst larger seeds usually contain toxins.

Tubers are more useful survival foods, and these, along with animal foods, should be the focus of survival. Most watercourses are home to at least one kind of tuber-bearing plant, and this habitat should always be searched first. The tubers of all rushes and sedges are edible raw. Watercourses are also rich in animal foods and fruits, and often lead the way to rescue.

In temperate Australia, open forests and woodlands often harbour tubers. Seek out bracken, orchids, lilies and murnong. In rainforest clearings look for long yams.

To dig tubers you will need a tool. A shovel is less efficient than a sturdy metal bar. Use a crowbar, jack handle, tent peg, knife, scrap metal, axe, or make a blunt-tipped digging stick from heavy wood, first hardening the tip in a fire. If possible, avoid tubers growing among rocks or tree roots.

If circumstances are truly desperate, it may be worth trying to leach the toxic tubers of round yams or Polynesian arrowroot. Grate the tubers against rough wood, stones or seashells, tie them inside a loose-weave cloth, and suspend this in a fast-flowing stream at least overnight, or for much longer if the water is slow-flowing, the cloth is tight-woven or the gratings are very coarse. Sample a small amount at first. The gratings should be edible raw but will taste better if wrapped in green grass and roasted over a fire. Alternatively, the tubers can be baked before grating and soaking; this makes them easier to grate. This technique is dangerous and should only be attempted under dire circumstances.

The inner hearts of palms, grasstrees and treeferns are very useful survival foods, although large amounts may cause digestive upset. The

Bush survival is easy wherever large colonies of tuber-bearing milkmaids (Burchardia umbellata) *occur.*

inner white leaf bases of many kinds of sedges can be nibbled raw. Fruits can be sampled at random and any that taste good will be safe to eat (but see the chapter on poisoning). Bat fruits are more likely to be edible than bird fruits (see Chapter 4, fruit section). Pale-coloured gums oozing from wattles and other trees, if palatable, can be eaten in small amounts, but may cause diarrhoea or constipation. Leaves and fungi should be largely avoided as they contribute few calories, and are not worth the risk of poisoning.

Animals are the ideal survival foods. Most are edible, and they contain more protein, fat and energy than plant foods. Try earthworms, moths, stick insects, cicadas, lizards, snakes (only the venom glands are dangerous), tortoises, and shellfish, which along seacoasts are the ideal staple foods. Turn over logs and rocks for termites (very nutritious), grubs and centipedes, and swat mosquitoes and march flies, which taste sweet. Grasshoppers are good eating unless they have been feeding on toxic plants, which give them an irritant taste. Avoid cane toads, green frogs, puffer fish, colourful caterpillars, stink bugs and wasps.

Cook wild foods where possible, but remember that the more dangerous toxins are not destroyed by heat.

Ignore the bushman's adage that plants with milky sap or red fruits are poisonous. Milky saps are often irritant, but the fruits or seeds produced by the same plant are not necessarily so, as wild figs, cultivated pawpaws and mangoes demonstrate. Edible fruits are more likely to be red than any other colour.

When tasting unknown plants, touch a portion to the lips, wait two minutes, then touch the tongue, wait two minutes, then chew a portion without swallowing, and wait two minutes, before finally eating a portion. This test is probably an unnecessary precaution in outback and southern Australia, but in the rainforests and in the coastal tropics it helps prevent poisoning from very irritant plants like the cunjevoi and related lilies with arrowhead-shaped leaves. Tasteless at first, they soon produce intense irritation and swelling in the mouth if chewed.

Be wary of very large seeds, avoid wild beans and peas, and take note of the poisoning chapter. These exceptions aside, learn to trust your sense of taste.

12

Changes

When English settlers first ventured onto Victoria's grassy plains they found a thriving Aboriginal society living on the sweet white roots of murnong—the yam daisy. Early travellers told of seeing "millions of murnong or yam, all over the plain", and noted that "the wheels of our dray used to turn them over by the bushel". The traveller G. A. Robinson saw women "spread over the plain as far as I could see them . . . I examined their bags and baskets on their return and each had a load as much as she could carry".

Life was easy on murnong. In 1886 near Echuca E. M. Curr found the tubers were "so abundant and so easily procured, that one might have collected in an hour, with a pointed stick, as many as would have served a family for the day".

But in the 1830s great herds of sheep and cattle were turned out onto the murnong fields. For the Aborigines it was the beginning of the end. The Goulburn Aborigine Moonin-Moonin lamented in 1839, only five years after the founding of Melbourne, that there were too many cattle and sheep, "plenty eat it murnong, all gone murnong".

By 1831, 700,000 sheep were tramping across Victoria. They ate not only the leaves of murnong, but rooted up the tubers as well. In 1845, only ten years after Victoria was settled, a Select Committee on Aborigines heard that the native foods had greatly declined. Grazing by stock had "rendered edible roots exceedingly scarce". F. Tuckfield noted sadly that "murnong and other valuable roots are eaten by the white man's sheep, and their deprivation, abuses and miseries are daily increasing".

The Aborigines faded away. Their food plant, once bountiful throughout Victoria's plains and open forests, now survives only in hilly woodlands where grazing is kept in check.

The tragedy of the Aborigines has been recounted many times, but the fate of their food plants is a forgotten story. The catastrophic decline of murnong was reconstructed recently by botanist Dr Beth Gott, from whose excellent article of 1983 this account was drawn. The sad story of many other food plants has never been told.

In a world grown weary of worthy issues, the plight of Australia's

The characteristic seedhead of the murnong (Microseris scapigera) *or yam daisy.*

rare plants is a last lost cause. Until Leigh, Boden and Briggs published *Extinct and Endangered Plants of Australia* in 1983, few people had ever entertained the idea of a rare Australian plant. We now know there are more than 70 extinct and several hundred endangered plants, nearly all of them threatened or destroyed by clearing or grazing.

Of the foods once gathered by Aborigines, the smaller plants grazed by rabbits and sheep and trampled by cattle and sheep have suffered the most. The tuber-producing murnong, lilies and orchids are especially vulnerable because they sprout leaves anew from their tubers each year. These short-lived leaves never develop the toughness of grass and shrub leaves, and they are eagerly sought out by cattle, rabbits and sheep. Of the 47 kinds of fringed lily, 20 are now rare, and grazing and clearing are blamed for their decline. Many ground orchids are also extremely rare.

Another group at risk are the peppercresses (*Lepidium*). Seven species are considered rare, four are presumed extinct. The spicy leaves are relished by cattle and rabbits—grazing animals from regions of the world where cresses are prominent in the flora. Cattle also trample these delicate plants.

Clearing has been the other major threat to Australian plants. The felling of the Big Scrub, a huge block of rainforest in northern New South Wales, pushed many plants close to extinction. The small-leaved tamarind (*Diploglottis campbellii*) survives as fewer than 20 wild trees. The durobby, the lilly pilly (*S. hodgkinsoniae*), the red bopple nut, an undescribed plum (*Davidsonia*) and a newly discovered quandong (*Elaeocarpus williamsianus*) are other rarities from this region.

But perversely, clearing has benefited many food plants. Native raspberries, grapes and stinging trees thrive along logging trails in rainforests. Coast beard heath, pigfaces and nitre bush abound on coastal dunes damaged by human activity. These plants are probably more common today than when Captain Cook charted the coasts. Even in the outback, food plants such as wild tomatoes, woollybutt and desert lime benefit from clearing and burning.

Unfortunately, the success of these plants is no consolation for the appalling degradation of Australia's rich ecosystems. Our surviving forest remnants have suffered a loss of diversity that is only now becoming recognised. The world of the Aborigines has passed away, and their loss is also ours.

Appendix I
Leaf Gallery

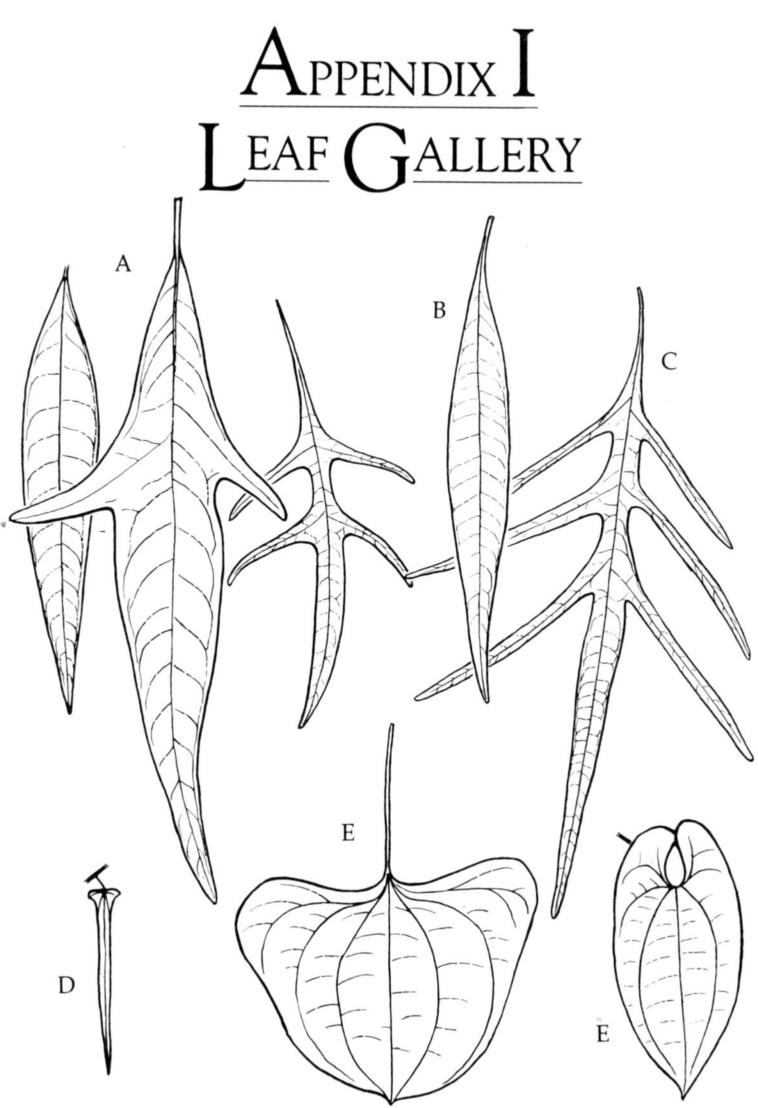

All leaves are shown at 40 per cent life size.

Kangaroo apples: A. *Solanum aviculare;* **B**. *S. laciniatum;* **C**. *S. vescum;* **Yams**: **D**. *Dioscorea hastifolia;* **E**. *D. transversa.*

F. Pepper tree (*Tasmannia insipida*); **G**. Mile-a-minute (*Ipomoea cairica*); **H**. Wild parsnip (*Trachymene incisa*); **Geebungs**: **I.** *Persoonia laurina*; **J**. *P. falcata*; **K**. *Persoonia* species; **L**. *P. cornifolia*.

M. Murnong (*Microseris scapigera*); **N**. Maloga bean (*Vigna lanceolata*); **O**. Konkerberry (*Carissa lanceolata*); **P**. Wild bauhinia tree (*Lysiphyllum gilvum*—leaf, flower and pod).

Appendix II
Tuber Gallery

Tubers can be useful guides to identification, especially among the ground orchids and lilies. They are often as distinctive as the above-ground features of the plant.

Ground lilies, orchids and murnong are difficult plants to find after their flowering season has ended—their presence is indicated only by a few withering leaves and a seed stalk. The dried seed stalk may persist for many months, and to the trained eye it points out the presence of hidden tubers. The seed stalks, like the tubers, are characteristic for each species, and some of these have been illustrated in the tuber gallery. Murnong's characteristic seedhead is illustrated on page 207.

Orchids, lilies, murnong, Polynesian arrowroot and most yams replace their tubers each year, and during the flowering season usually bear tubers in pairs, representing the past and present year's crops. The newer tubers, distinguished by their paler colour and crisper texture, are nearly always the more palatable. The older tubers shrivel and darken as the plant empties them of nutrients. Aboriginal women knew the best time to harvest each kind of tuberous plant, which was usually after the flowers, leaves and old tubers had withered away.

Orchids, lilies, murnong and nalgoo are becoming uncommon plants, and their tubers should not be exploited except under dire circumstances. The drawings on the following pages are about 40 per cent of life size.

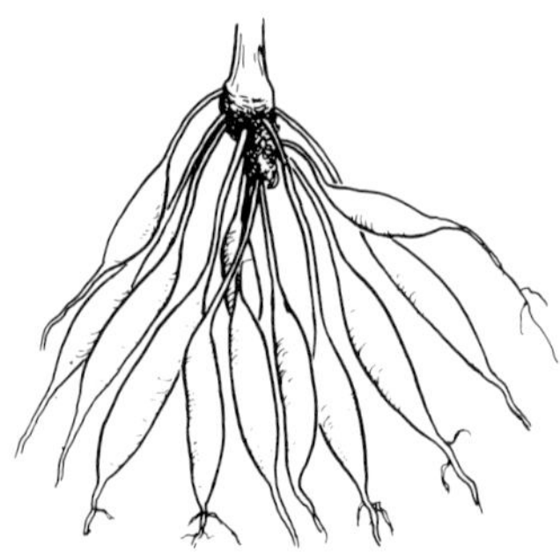

Tubers of pale vanilla lily (*Arthropodium milleflorum*).

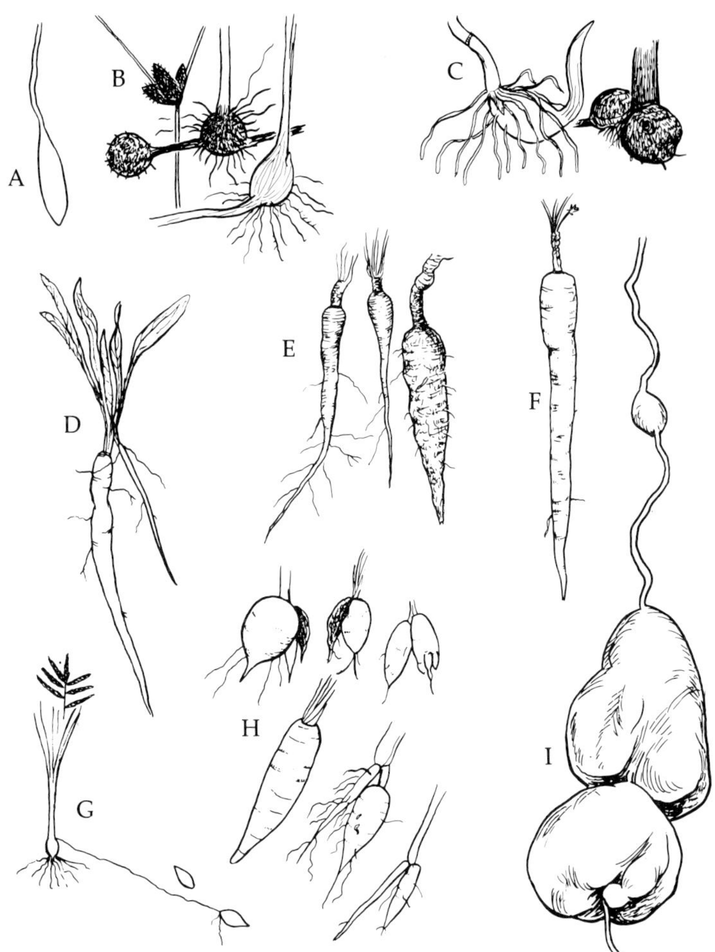

A. Water ribbons *(Triglochin species)*: juicy, fibrous; **B.** Sea club-rush *(Bolboschoenus caldwellii*, seed spikes, old and new tubers): sweet fibrous; **C.** Marsh club-rush *(B. fluviatilis*, new and old tubers): sweet, fibrous; **D.** Murnong *(Microseris scapigera*, sub-alpine form): starchy, bitter; **E.** cranesbills *(Geranium)*: astringent, unpalatable; **F.** Wild parsnip *(Trachymene incisa)*: starchy, aromatic; **G.** nalgoo *(Cyperus bulbosus*, seed spike and tubers): pleasantly starchy; **H.** murnong *(Microseris scapigera*, lowland form): starchy, bitter or sweet; **I.** Bush banana *(Marsdenia viridiflora)*: juicy, insipid.

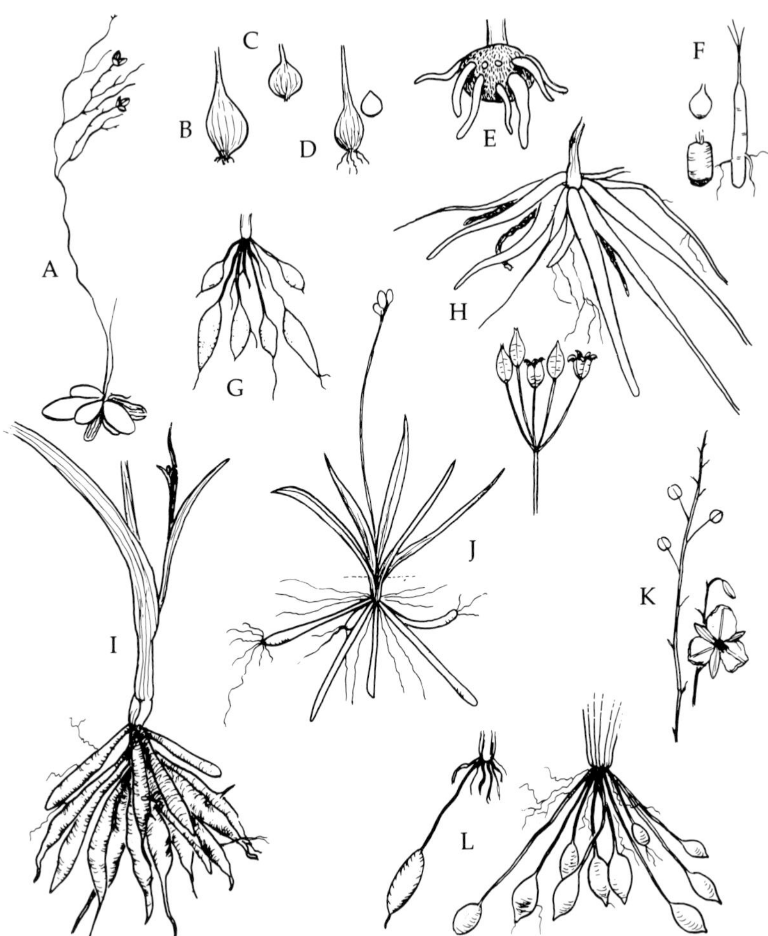

A. Twining fringed lily (*Thysanotus patersonii*): watery, bittersweet; **B**. Early nancy (*Wurmbea centralis*): starchy, bitter aftertaste; **C**. *W. biglandulosa*: bitter, unpalatable; **D**. *W. dioica*: bitter, unpalatable; **E**. Bulbine lily (*Bulbine bulbosa*): glutinous, bland; **F**. Golden stars (*Hypoxis* species): starchy, often bitter; **G**. Fringed lily (*Thysanotus tuberosus*): watery, sweet; **H**. Milkmaids (*Burchardia umbellata*, tubers and seed pods): crisp, starchy; **I**. Pink swamp lily (*Murdannia graminea*): fibrous, insipid; **J**. Blue star (*Chamaescilla corymbosa*): pleasantly starchy; **K**. common chocolate lily (*Dichopogon strictus*, stalks and tubers): watery, bittersweet; **L**. Pale grass lily (*Caesia calliantha*).

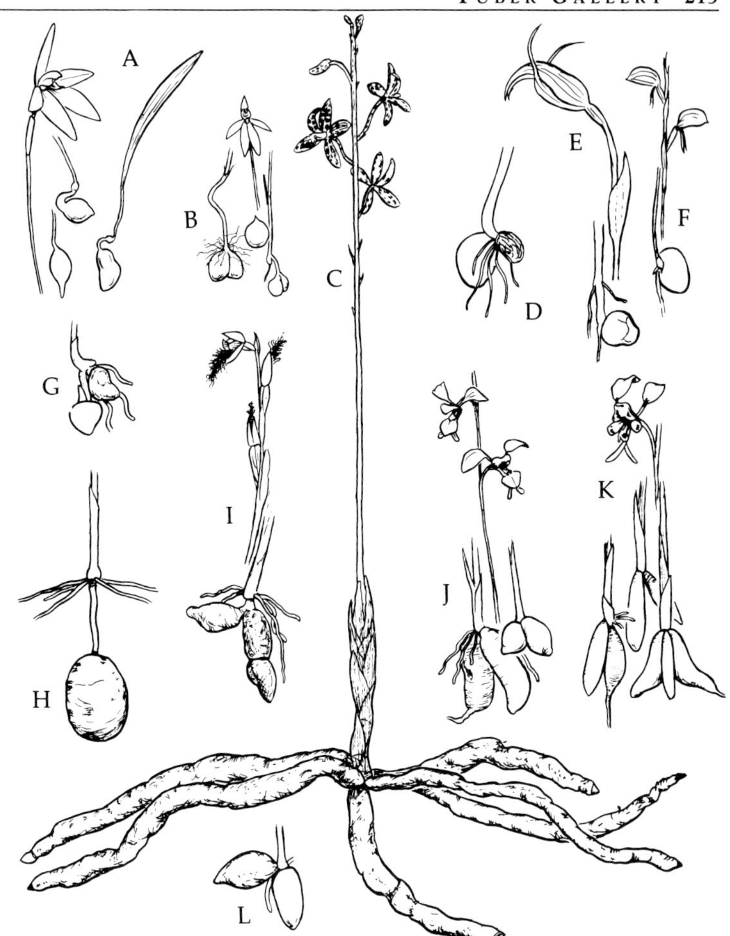

A. Common wax-lip (*Glossodia major*): watery, bittersweet; **B**. Pink fingers (*Caladenia carnea*): watery, sweetish; **C**. Hyacinth orchid (*Dipodium punctatum*): watery, fibrous; **D**. Onion orchid (*Microtis*): bland, starchy; **E**. Greenhood (*Pterostylis*): watery, sweetish; **F**. Tall greenhood (*P. longifolia*): watery, bitter; **G**. Alpine leek orchid (*Prasophyllum alpinum*): juicy, bittersweet; **H**. Brown-beaks (*Lyperanthus suaveolens*): juicy, fragrant; **I**. Bearded orchid (*Calochilus*): starchy, bitter; **J**. Golden donkey orchid (*Diuris aurea*): bland, glutinous; **K**. Leopard orchid (*D. maculata*): bland, very glutinous; **L**. Spotted sun orchid (*Thelymitra ixioides*): bland, glutinous.

REFERENCES

Altman, J. C. (1984) The Dietary Utilisation of Flora and Fauna by Contemporary Hunter-gatherers at Momega Outstation, North-central Arnhem Land. **Australian Aboriginal Studies** 1984(1). 35–46.

Angas, G. F. (1847) **Savage Life and Scenes in Australia and New Zealand.** Smith, Elder, London.

Backhouse, J. (1843) **A Narative of a Visit to the Australian Colonies.** Hamilton Adams, London.

Bancroft, J. (1885) Food of the Aborigines of Central Australia. **Proceedings of the Royal Society of Queensland** 1. 104–107.

Banks, J. (1962) **The Endeavour Journal of Joseph Banks 1768–1771.** J. C. Beagehole (ed.). Angus & Robertson, Sydney.

Beaton, J. M. (1977) Dangerous Harvest. Ph.D. thesis on cycads. Australian National University, Canberra.

Brand, J. C. and V. Cherikoff (1985) Australian Aboriginal Bushfoods: The Nutritional Composition of Plants from Arid and Semi-arid Areas. **Australian Aboriginal Studies** 1985(2). 38–46.

Brand, J. C., V. Cherikoff and A. S. Truswell (1985) The Nutritional Composition of Australian Aboriginal Bushfoods. 3. Seeds and Nuts. **Food Technology in Australia** 37. 275–279.

Brand, J. C., C. Rae, J. McDonnell, A. Lee, V. Cherikoff and A. S. Truswell (1983) The Nutritional Composition of Australian Aboriginal Bushfoods 1. **Food Technology in Australia** 35. 293–298

Brown, G. R. (1894) Notes on the Uses etc. of some Native Plants in the Port Macquarie District. **Agricultural Gazette of New South Wales** 4(9). 680–682.

Cane, S., J. Stockton and A. Vallance (1979) A Note on the Diet of Tasmanian Aborigines. **Australian Archaeology** 9. 77–81.

Cherikoff, V., J. C. Brand and A. S. Truswell (1985) The Nutritional Composition of Australian Aboriginal Bushfoods. 2. Animal Foods. **Food Technology in Australia** 37(5). 208–211.

Cherikoff, V. and J. Isaacs (1989) **The Bush Food Handbook.** Ti Tree Press, Sydney.

Cleland, J. B. (1966) The Ecology of the Aboriginal in South and Central Australia (in) B. C. Cotton (ed.) **Aboriginal Man in South and Central Australia. Part 1.** Government Printer, Adelaide.

Cleland, J. B. and T. H. Johnston (1939) Aboriginal Names and Uses of Plants in the Northern Flinders Ranges. **Transactions of the Royal Society of South Australia** 63(2). 172–179.

Cleland, J. B. and N. B. Tindale (1959) The Native Names and Uses of Plants at Haast Bluff, Central Australia. **Transactions of the Royal Society of South Australia** 82. 123–140.

Collins, D. (1798) **An Account of the English Colony in New South Wales.** Cadell & Davies, London.

Crawford, I. M. (1982) **Traditional Aboriginal Plant Resources in the Kalumburu Area.** Western Australian Museum, Perth.

Cribb, A. B. and J. W. Cribb (1974) **Wild Food in Australia.** Collins, Sydney.

Cribb, A.B. and J.W. Cribb (1985) **Plantlife of the Great Barrier Reef**. University of Queensland Press, Brisbane.

Cunningham, P. (1827) **Two Years in New South Wales.** H. Colburn, London.

Dawson, J. (1881) **Australian Aborigines.** George Robertson, Melbourne.

Duncan-Kemp, A. (1933) **Our Sandhill Country.** Angus & Robertson, Sydney.

Gandevia, B. (1981) A-going for Greens (in) D. J. Carr and S. G. M. Carr (eds) **Plants and Man in Australia.** Academic Press, Sydney.

George, A. S. (1984) **The Banksia Book.** Kangaroo Press, Sydney.

Gott, B. (1982) Ecology of Root Use by Aborigines of Southern Australia. **Archaeology in Oceania** 17. 59–67.

Gott, B. (1983) Murnong—**Microseris scapigera**: a study of a staple food of Victorian Aborigines. **Australian Aboriginal Studies** 1983(2). 2–18.

Grey, G. (1841) **Journals of Two Expeditions of Discovery in North-west and Western Australia during the years 1837, 38, and 39.** 2 vols. T. & W. Boone, London.

Harris, D. R. (1977) Subsistence Strategies across Torres Strait (in) J. Allen, J. Golsen and R. Jones (eds) **Sunda and Sahul.** Academic Press, London.

Hassell, E. (1975) **My Dusky Friends.** C. W. Hassell, Perth.

Hyland, B. P. M. (1983) A Revision of *Syzygium* and Allied Genera (Myrtaceae) in Australia. **Australian Journal of Botany**. Supplementary Series 9.

Hynes, R. A. and A. K. Chase (1982) Plants, Sites and Domiculture: Aboriginal Influence upon Plant Communities in Cape York Peninsula. **Archaeology in Oceania** 17(1). 38–50.

Isaacs, J. (1987) **Bush Food.** Weldons, Sydney.

James, K. W. (1983) Analysis of indigenous Australian foods. **Food Technology in Australia** 35. 342–343.

Johnston, T. H. and J. B. Cleland (1943) Native Names and Uses of Plants in the North-eastern Corner of South Australia. **Transactions of the Royal Society of South Australia** 66(1). 149–173

Koch, M. (1898) A List of Plants collected on Mt Lyndhurst Run, South Australia. **Transactions of the Royal Society of South Australia** 22. 101–118.

Labillardiere, J. (1800) **Voyage in Search of La Perouse.** Stockdale, London.

Latz, P. K. (1982) **Bushfires and Bushtucker: Aborigines and Plants in Central Australia.** M.A. Thesis, University of New England.

Leichhardt, L. (1847) **Journal of an Overland Expedition in Australia, from Moreton Bay to Port Essington, a Distance of Upwards of 3000 miles, During the Years 1844–1845**. T. & W. Boone, London.

Leigh, J., R. Boden and J. Briggs (1984) **Extinct and Endangered Plants of Australia.** Macmillan, Melbourne.

Levitt, D. (1981) **Plants and People: Aboriginal Uses of Plants on Groote Eylandt.** Australian Institute of Aboriginal Studies, Canberra.

Low, T. (1985) **Wild Herbs of Australia & New Zealand.** Angus & Robertson, Sydney.

Low, T. (1987) Explorers and Poisonous Plants (in) J. Covacevich, P. Davie and J. Pearn (eds) **Toxic Plants & Animals: A Guide for Australia.** Queensland Museum, Brisbane.

Low, T. (1988) **Wild Food Plants of Australia.** Angus & Robertson, Sydney.

Low, T. (1989) **Bush Tucker.** Angus & Robertson, Sydney.

Mann, D. D. (1811) **The Present Picture of New South Wales.** Booth, London.

Maiden, J. H. (1889) **The Useful Native Plants of Australia (including Tasmania).** Trubner and Co., London.

Maiden, J. H. (1898) Some Plant-foods of the Aborigines. **Agricultural Gazette of New South Wales** 9(4). 349–354.

Maiden, J. H. (1898) A New Indigenous Food-plant. **Agricultural Gazette of New South Wales** 9(4). 355

Maiden, J. H. (1900) Native Food Plants. **Agricultural Gazette of New South Wales** 10(2). 115–130; 10(4). 279–290; 10(7). 618–629; 10(8). 730–740.

Meagher, S. J. (1974) The Food Resources of the Aborigines of the South-west of Western Australia. **Records of the Western Australian Museum** 3. 14–65.

Mitchell, T. L. (1838) **Three Expeditions into the Interior of Eastern Australia, with Descriptions of the Recently Explored Region of Australia Felix, and the Present Colony of New South Wales.** Boone, London.

Mitchell, T. L. (1848) **Journal of an Expedition into the Interior of Tropical Australia, in Search of a Route from Sydney to the Gulf of Carpentaria.** Longman, Brown, Green, and Longmans, London.

Mueller, F. (1867) Documents relating to the Intercolonial Exhibition of 1866–67. Blundell & Co., Melbourne.

Nixon, F. R. (1857) **The Cruise of the Beacon.** Bell & Daldy, London.

O'Connell, J. F., P. K. Latz and P. Barnett (1983) Traditional and Modern Plant Use Among the Alyawara of Central Australia. **Economic Botany** 37. 80–109.

Palmer, E. (1884) On Plants used by the Natives of North Queensland, Flinders and Mitchell Rivers, for Food, Medicine, &c., &c. **Journal and Proceedings of the Royal Society of New South Wales** 17. 93–113.

Pate, J. S. and K. W. Dixon (1982) **Tuberous, Cormous and Bulbous Plants.** University of Western Australia Press, Nedlands.

Petrie, C. C. (1904) **Tom Petrie's Reminiscences of Early Queensland.** Watson, Ferguson & Co., Brisbane.

Phillip, A. (1789) **The Voyage of Governor Phillip to Botany Bay.** Stockdale, London.

Pijl, L. van der (1969) **Principles of Dispersal in Higher Plants.** Springer-Verlag, Berlin.

Rawson, L. (1984) **Australian Enquiry Book.** Kangaroo Press, Sydney.

Roth, H. L. (1890) **The Aborigines of Tasmania.** F. King & Sons, Halifax.

Roth, W. E. (1901) Food: Its search, Capture and Preparation. **North Queensland Ethnography** 3. 1-31.

Roth, W. E. (1903) Notes of Savage Life in the Early Days of West Australian Settlement. **Proceedings of the Royal Society of Queensland.** 17. 45-69.

Schurmann, C. W. (1879) in **The Native Tribes of South Australia.** E. S. Wigg & Son, Adelaide.

Smith, M. and A. C. Kalotas (1985) Bardi Plants: An Annotated List of Plants and Their Use by the Bardi Aborigines of Dampierland, in North-western Australia. **Records of the Western Australian Museum** 12(3). 317-359.

Smyth, R. B. (1878) **Aborigines of Victoria.** Government Printer, Melbourne.

Stephens, E. (1889) The Aborigines of Australia. **Journal of the Royal Society of New South Wales** 23. 476-502.

Stuart, J. M. (1865) **Explorations in Australia: The Journals of John McDouall Stuart.** Saunders, Otley and Co, London.

Thozet, A. (1886) **Catalogue of some of the Roots, Tubers, Bulbs and Fruits, used as Vegetable Food by the Aboriginals of Northern Queensland, Australia.** W. H. Buzzacott, Rockhampton.

Turner, F. (1905) Botany of North-western New South Wales. **Proceedings of the Linnean Society of New South Wales** 30. 32-90.

Welsby, T. (1922) **Memories of Amity.** Watson, Ferguson & Co., Brisbane.

Wilhelmi, C. (1860) Manners and Customs of the Australian Natives, in Particular of the Port Lincoln District. **Transactions of the Royal Society of Victoria** 5. 164-203.

Woolls, W. (1867) Indigenous Vegetables (in) **A Contribution to the Flora of Australia.** F. White, Sydney.

PHOTOGRAPHIC CREDITS

T.J. Hawkeswood & T. Low: 26-29, 46-47, 53-54, 58, 60-64, 66-67, 71-73, 77, 82, 87-92, 99-103, 106, 108, 110-113, 122-123, 126-130, 135-136, 154, 162, 164, 167-8, 170, 174, 182.
Jeanette Covacevich: 57.
Bruce Cowell: 78.
Murray Fagg: 70.
Owen Foley: 96.
Peter Latz: 163.
Robert Rankin: 85, 141.
Queensland Herbarium: 138, 177.
All other photos are by the author.

INDEX

*Numbers in **bold** italics indicate the page where the plant is described and illustrated.*

A

C

G

H

I

J

K

L

M

N

O

P

S

T

X

Y

Z